Chee Peng Lim, Lakhmi C. Jain, and Satchidananda Dehuri (Eds.)

Innovations in Swarm Intelligence

Studies in Computational Intelligence, Volume 248

Editor-in-Chief
Prof. Janusz Kacprzyk
Systems Research Institute
Polish Academy of Sciences
ul. Newelska 6
01-447 Warsaw
Poland
E-mail: kacprzyk@ibspan.waw.pl

Further volumes of this series can be found on our homepage: springer.com

Vol. 228. Lidia Ogiela and Marek R. Ogiela
Cognitive Techniques in Visual Data Interpretation, 2009
ISBN 978-3-642-02692-8

Vol. 229. Giovanna Castellano, Lakhmi C. Jain, and Anna Maria Fanelli (Eds.)
Web Personalization in Intelligent Environments, 2009
ISBN 978-3-642-02793-2

Vol. 230. Uday K. Chakraborty (Ed.)
Computational Intelligence in Flow Shop and Job Shop Scheduling, 2009
ISBN 978-3-642-02835-9

Vol. 231. Mislav Grgic, Kresimir Delac, and Mohammed Ghanbari (Eds.)
Recent Advances in Multimedia Signal Processing and Communications, 2009
ISBN 978-3-642-02899-1

Vol. 232. Feng-Hsing Wang, Jeng-Shyang Pan, and Lakhmi C. Jain
Innovations in Digital Watermarking Techniques, 2009
ISBN 978-3-642-03186-1

Vol. 233. Takayuki Ito, Minjie Zhang, Valentin Robu, Shaheen Fatima, and Tokuro Matsuo (Eds.)
Advances in Agent-Based Complex Automated Negotiations, 2009
ISBN 978-3-642-03189-2

Vol. 234. Aruna Chakraborty and Amit Konar
Emotional Intelligence, 2009
ISBN 978-3-540-68606-4

Vol. 235. Reiner Onken and Axel Schulte
System-Ergonomic Design of Cognitive Automation, 2009
ISBN 978-3-642-03134-2

Vol. 236. Natalio Krasnogor, Belén Melián-Batista, José A. Moreno-Pérez, J. Marcos Moreno-Vega, and David Pelta (Eds.)
Nature Inspired Cooperative Strategies for Optimization (NICSO 2008), 2009
ISBN 978-3-642-03210-3

Vol. 237. George A. Papadopoulos and Costin Badica (Eds.)
Intelligent Distributed Computing III, 2009
ISBN 978-3-642-03213-4

Vol. 238. Li Niu, Jie Lu, and Guangquan Zhang
Cognition-Driven Decision Support for Business Intelligence, 2009
ISBN 978-3-642-03207-3

Vol. 239. Zong Woo Geem (Ed.)
Harmony Search Algorithms for Structural Design Optimization, 2009
ISBN 978-3-642-03449-7

Vol. 240. Dimitri Plemenos and Georgios Miaoulis (Eds.)
Intelligent Computer Graphics 2009, 2009
ISBN 978-3-642-03451-0

Vol. 241. János Fodor and Janusz Kacprzyk (Eds.)
Aspects of Soft Computing, Intelligent Robotics and Control, 2009
ISBN 978-3-642-03632-3

Vol. 242. Carlos A. Coello Coello, Satchidananda Dehuri, and Susmita Ghosh (Eds.)
Swarm Intelligence for Multi-objective Problems in Data Mining, 2009
ISBN 978-3-642-03624-8

Vol. 243. Imre J. Rudas, János Fodor, and Janusz Kacprzyk (Eds.)
Towards Intelligent Engineering and Information Technology, 2009
ISBN 978-3-642-03736-8

Vol. 244. Ngoc Thanh Nguyen, Rados law Piotr Katarzyniak, and Adam Janiak (Eds.)
New Challenges in Computational Collective Intelligence, 2009
ISBN 978-3-642-03957-7

Vol. 245. Oleg Okun and Giorgio Valentini (Eds.)
Applications of Supervised and Unsupervised Ensemble Methods, 2009
ISBN 978-3-642-03998-0

Vol. 246. Thanasis Daradoumis, Santi Caballé, Joan Manuel Marquès, and Fatos Xhafa (Eds.)
Intelligent Collaborative e-Learning Systems and Applications, 2009
ISBN 978-3-642-04000-9

Vol. 247. Monica Bianchini, Marco Maggini, Franco Scarselli, and Lakhmi C. Jain (Eds.)
Innovations in Neural Information Paradigms and Applications, 2009
ISBN 978-3-642-04002-3

Vol. 248. Chee Peng Lim, Lakhmi C. Jain, and Satchidananda Dehuri (Eds.)
Innovations in Swarm Intelligence, 2009
ISBN 978-3-642-04224-9

Chee Peng Lim, Lakhmi C. Jain,
and Satchidananda Dehuri (Eds.)

Innovations in Swarm Intelligence

Springer

Dr. Chee Peng Lim
University of South Australia
Adelaide
The Mawson Lakes SA 5095
Australia

Prof. Lakhmi C. Jain
University of South Australia
Adelaide
The Mawson Lakes SA 5095
Australia
E-mail: Lakhmi.jain@unisa.edu.au

Dr. Satchidananda Dehuri
Soft Computing Laboratory
Department of computer Science
Yonsei University
262 seongsanro
sudaemoon-gu, Seoul 120-749
Korea

ISBN 978-3-642-26056-8 e-ISBN 978-3-642-04225-6

DOI 10.1007/978-3-642-04225-6

Studies in Computational Intelligence ISSN 1860-949X

Softcover reprint of the hardcover 1st edition 2010

Typeset & Cover Design: Scientific Publishing Services Pvt. Ltd., Chennai, India.

Printed in acid-free paper

9 8 7 6 5 4 3 2 1

springer.com

Preface

In recent years, swarm intelligence has attracted a lot of researchers' attention. A variety of swarm intelligence models have been proposed and applied successfully to solve many real-world problems. In general, the concept of swarm intelligence is inspired by the social behaviour of gregarious insects and other animals. The emergent behaviour of multiple unsophisticated agents interacting among themselves and with their environment leads to a functional strategy that is useful to achieve complex goals in an efficient manner. For instance, ants, which are almost blind to their environment, are capable of finding the shortest path from their colony to their food sources and back. Bees perform waggle dances to convey useful information on nectar sources to their hive mates. A number of desirable properties also exist in swarm intelligence models, which include feedback and adaptation to changing environments, and multiple decentralized interactions among agents to work collaboratively as a group in completing complex tasks.

From the computational point of view, swarm intelligence models are largely stochastic search algorithms. They are useful for undertaking distributed and multimodal optimization problems. The search process is robust and efficient in maintaining diversity. A mechanism to impose a form of forgetting is also adopted in some swarm intelligence algorithms such that the solution space can be explored in a comprehensive manner. Thus, the algorithms are able to avoid convergence to a locally optimal solution, and, at the same time, to arrive at a global optimized solution with a high probability.

In this research book, a small collection of recent innovations in swarm intelligence is presented. The swarm intelligence techniques covered include particle swarm optimization and hybrid methods, ant colony optimization and hybrid methods, bee colony optimization, glowworm swarm optimization, and complex social swarms. Applications of swarm intelligence to operational planning of energy plants, modelling and control of nanorobots, classification of documents, identification of disease biomarkers, and prediction of gene signals are described. The book is useful to researchers, practising professionals, and undergraduate as well as graduate students of all disciplines.

The editors would like to express their utmost gratitude and appreciation to the authors for their contributions. The editors are grateful to the reviewers for their constructive comments and suggestions. Thanks are also due to the excellent editorial assistance by staff at Springer-Verlag and SCI Data Processing Team of Scientific Publishing Services.

C.P. Lim
L.C. Jain
S. Dehuri

Table of Contents

1

Advances in Swarm Intelligence

Chee Peng Lim and Lakhmi C. Jain

School of Electrical & Information Engineering
University of South Australia, Australia

Abstract. In this chapter, advances in techniques and applications of swarm intelligence are presented. An overview of different swarm intelligence models is described. The dynamics of each swarm intelligence model and the associated characteristics in solving optimization as well as other problems are explained. The application and implementation of swarm intelligence in a variety of different domains are discussed. The contribution of each chapter included in this book is also highlighted.

1 Introduction

In recent years, swarm intelligence, which can be considered as a branch of Artificial Intelligence techniques, has attracted much attention of researchers, and has been applied successfully to solve a variety of problems. A swarm can be viewed as a group of agents cooperating with certain behavioural pattern to achieve some goal [1]. There are a number of different models of swarm intelligence that have been proposed and investigated, and among the most commonly used swarm intelligence models include ant colony optimization [2], [3], particle swarm optimization [4], [5], honey bee swarming [6] [7], and bacterial foraging [8] [9].

In general, swarm intelligence deals with modelling of the collective behaviours of simple agents interacting locally among themselves, and their environment, which leads to the emergence of a coherent functional global pattern [10]. These models are inspired by the social behaviour of insects and other animals [11]. From the computational point of view, swarm intelligence models are computing algorithms that are useful for undertaking distributed optimization problems. The fundamental principle of swarm intelligence hinges on probabilistic-based search algorithms. All swarm intelligence models exhibit a number of general properties [12]. Each entity of the swarm is made of a simple agent. Communication among agents is generally indirect and short. Cooperation among agents is realised in a distributed manner without a centralised control mechanism. These properties make swarm intelligence models easy to be realised and extended, such that a high degree of robustness can be achieved. In other words, the entire swarm intelligence model is simple in nature. However, the collective colony-level behaviour of the swarm that emerges out of the interactions is useful in achieving complex goals [13].

The organisation of this chapter is as follows. In the next section, different types of swarm intelligence models are described. Then, application and implementation of swarm intelligence models are discussed in section 3. In section 4, the contribution of

C.P. Lim et al. (Eds.): Innovations in Swarm Intelligence, SCI 248, pp. 1–7.
springerlink.com

each chapter in the book is highlighted. A summary of concluding remarks is presented in section 5.

2 Swarm Intelligence Models

According to [14], there are five basic principles of swarm intelligence. They are (i) proximity: the ability to perform simple computation of time and space so as to respond to environmental stimuli; (ii) quality: the ability to respond to quality factors, e.g. food and safety; (iii) diverse response: the ability to distribute resources and to safeguard against environmental changes; (iv) stability: the ability to maintain the group behaviour against every fluctuation in the environment; (v) adaptibility: the ability to change the group behaviour that leads to better adaptation to the environment. As such, it is necessary to strike a balance between the principles of stability and adaptability [14].

In the following section, three commonly used swarm intelligence models are described. They are ant colony optimization, particle swarm optimization, and bee colony optimization.

2.1 Ant Colony Optimization

Ant colony optimization was proposed by Dorigo [2]. It was inspired by the social behaviour and routing technique of ants in search of food. Ants are able to establish the shortest path from their colony to their food sources and back. When searching for food, ants first wander around their environment randomly. Upon finding food, they return to their colony while laying a trail of chemical substance called pheromone along their path. Essentially, pheromone is used for communication purposes. Other ants will detect the trail of pheromone and follow the path. In the process, more pheromone is deposited along the path. The richer the trail of pheromone along a path, the more likely other ants will detect and follow the path. In other words, ants tend to choose the paths marked by the strongest pheromone concentration. However, the pheromone will evaporate over time, therefore reducing its attractive strength. The longer the time ants take to transverse a path from their colony to the food source and back, the quicker the pheromone will evaporate. In other words, the pheromone along a shorter path will be reinforced quicker because when other ants follow the path, they keep adding their pheromone and making the pheromone deposit stronger before it evaporates. As a result, the shortest path between the ant colony and the food source will emerge eventually.

From the computational point of view, one of the advantages of pheromone evaporation is to avoid convergence to a locally optimal solution [9]. Indeed, pheromone evaporation is a useful mechanism to impose a form of forgetting, hence allowing the solution space to be explored in a comprehensive manner. Variants of ant colony optimization have been proposed, e.g. ant colony system [15] and max-min ant system [16].

2.2 Particle Swarm Optimization

Inspired by the social behaviour of bird flocking and fish schooling, particle swarm optimization is an evolutionary computation model that has its roots in artificial life.

It was first proposed by Eberhart and Kennedy [4]. Unlike the genetic algorithm, particle swarm optimization does not utilise filtering operators such as crossover and mutation in searching for solutions. Instead, it utilizes a population of particles that "fly" through a multi-dimensional search space with given velocities. Each particle encodes a single intersection of all search dimensions. The associated position and velocity of each particle are randomly generated. At each generation, the velocity of the particle is stochastically adjusted according to the historical best position for the particle itself and the neighbourhood best position. This is accomplished by using some fitness evaluation function. The movement of each particle evolves to an optimal or near-optimal solution.

Particle swarm optimization model has several advantages. The search mechanism is robust and efficient in maintaining diversity, and is able to arrive at the global optimization solution with a high probability. The algorithm is easy to realize with only a few adjustable parameters, is fast converging, and can be conducted with parallel computation. Variants of particle swarm optimization have been proposed, e.g. canonical particle swarm [17] and fully-informed particle swarm [18].

2.3 Bee Colony Optimization

A behavioural model of self-organization for a colony of honey bees was proposed by Seeley [6]. It was inspired by the foraging behaviours of bees. Foraging bees are sent out from their colony to search for promising food sources (flower patches). Upon finding a good food source, a foraging bee returns to the hive and informs its hive mates via a waggle dance. The mystery of bee waggle dances was decoded by von Frisch [19]. In essence, the waggle dance is a communication tool with other bees. The foraging bee conveys three pieces of important information to other bees through the waggle dance, i.e., the distance, the direction, and the quality of the food source. In particular, the foraging bee uses this waggle dance as a means to convince other bees to be followers and to go back to the food source. It is believed that the recruitment of bees is a function of the quality of the food source. As a result, more bees are attracted to more promising food sources. Such a strategy is deployed by the bee colony to obtain quality food in a fast and efficient manner.

The bee foraging behaviours exhibit the characteristics of self-organization, adaptiveness, and robustness. There are several advantages of bee colony optimization from the computational viewpoint, i.e., it avoids locally optimal solutions, it searches for the best solution based on the solutions obtained by the entire bee colony, and it is adaptive to changes in the environment [20]. Variants of bee colony optimization have been proposed, e.g. virtual bee algorithms [21] and fuzzy bee system [7].

3 Application and Implementation of Swarm Intelligence

In this section, a small fraction of the applications of the three swarm intelligence models discussed in section 2 is presented. In addition, hardware realization of two popular swarm intelligence models, i.e., ant colony optimization and particle swarm optimization, is described.

Ant colony optimization and its variants have been applied to tackle various optimization problems. As listed in [13], application areas of ant colony optimization-based models include travelling salesman problem, network routing, graph colouring, shortest common supersequence, quadratic assignment, machine scheduling, vehicle routing, multiple knapsack, frequency assignment, as well as sequential ordering.

In the power systems area, particle swarm optimization has been applied to tackle a wide variety of problems. A comprehensive review on variants of particle swarm optimization and their applicability to various power systems related problems is presented in [5], e.g. reactive power and voltage control, economic dispatch, power system reliability and security, generation expansion problem, state estimation, load flow and optimal power flow, and power system identification and control.

In the area of bioinformatics, swarm intelligence models have started to attract attention from researchers. A survey presented in [22] reveals that ant colony optimization and particle swam optimization techniques are useful for undertaking a variety of problems in bioinformatics, e.g. clustering of gene expression data, molecular docking problem, multiple sequence alignment problem, construction of phylogenetic trees, RNA secondary structure prediction, protein secondary structure prediction, and fragment assembly problem.

Bee colony optimization has been used to undertake complex transportation tasks, e.g., the Ride-Matching problem [7]. Other applications discussed in [23, 24, 25] include travelling salesman problem, routing, and wavelength assignment in optical networks, job shop scheduling. Methods for solving continuous optimization problems using bee colony optimization are discussed in [26]. A table listing a variety of problems tackled using bee colony optimization-based methods, which include data mining, vehicle routing, telecommunication network routing, water resources management problems, economic power dispatch, is presented in [26].

In terms of hardware realization, implementation of ant colony optimization in evolvable hardware has been reported. Among the progress made include colony optimization-based random number generator, field programmable gate arrays, digital circuits, as well as infinite impulse response filters [27]. Nevertheless, challenges that relate to online realization, robustness, generalization, as well as the disaster problem due to increasing complexity of optimization problems, need to be addressed in future implementation of ant colony optimization-based evolvable hardware [27]. On the other hand, particle swarm optimization-based hardware for neural network training, controller design and tuning, mobile robots, fault-tolerance sensor systems, as well as ant colony optimization-based hardware for wireless sensor networks is described in [28].

4 Chapters Included in This Book

The book contains twelve chapters. Chapter one presents an introduction to swarm intelligence, which include various swarm intelligence models and their applications. Chapter two presents a review on the most representative particle swarm optimization-based approaches that are used to deal with multimodal optimization problems. A case study on searching for solutions of nonlinear equations is also provided. The basic principles of bee colony optimization are presented in Chapter three. A survey on algorithms inspired by bees' behaviours is included. Applications of bee colony optimization to modelling of complex engineering and management problems, which include the travelling salesman problem, ride-matching, routing, and scheduling, are described. A novel swarm intelligence technique known as glowworm swarm optimization for simultaneous searching of multiple optima of multimodal functions is described in Chapter four. Benchmark functions are used to analyse the performance of glowworm swarm optimization.

In Chapter five, the effects of social influence on specialisation, i.e., division of labour, in complex agent systems connected via social networks are analysed. Simulations to investigate the sensitivity of the social influence rate on the overall level of system specialization are conducted. A review on the issue of computational complexity of ant colony optimization is presented in Chapter six. Hybridization of ant colony optimization with local search is discussed, and examples to demonstrate the effectiveness of the hybrid model are demonstrated. In Chapter seven, a hybrid genetic algorithm and particle swarm optimization model are presented. The hybrid model aims to promote global exploration and local exploitation of solutions. The model is tested on two multi-resolution parameter optimization problems. Application of particle swarm optimization and ant colony optimization in a multi-agent system architecture is described in Chapter eight. The proposed approach is used to predict translational initiation sites of gene signals by using benchmark data sets.

Application of particle swarm optimization-based methods to energy plants is presented in Chapter nine. An optimal operational planning and control system based on particle swarm optimization is described. In Chapter ten, a pilot study on a nanorobot swarm system is described. Use of the perceptive particle swarm optimization algorithm for nanorobot control is discussed, and application of the nanorobot swarm system to surface coating is demonstrated. In Chapter eleven, an ant colony optimization algorithm for document classification is described. By using the ant colony optimization module, a thematic word map is created to assist in the representation of documents in the pattern space. The performance of the ant colony optimization-based system is compared with that of the self-organizing map neural network. Methods to identify a set of representative features from social communities of genes are examined in Chapter twelve. Application of the methods to microarray data classification is included.

5 Summary

In this chapter, an overview on recent advances in swarm intelligence models has been presented. The characteristics and basic principles of swarm intelligence have been explained. Three popular swarm intelligence models, i.e., ant colony optimization, particle swarm optimization, and bee colony optimization, have been described. In addition, a small fraction of a wide variety of application areas of the three swarm intelligence models has been discussed. Issues related to hardware implementation and realization of swarm intelligence models have also been described. Finally, the contributions of each chapter in this book have been highlighted.

References

[1] Grosan, C., Abraham, A., Monica, C.: Swarm Intelligence in Data Mining. In: Abraham, A., Grosan, C., Ramos, V. (eds.) Swarm Intelligence in Data Mining. SCI, vol. 34, pp. 1–16. Springer, Heidelberg (2006)

[2] Colorni, A., Dorigo, M., Maniezzo, V.: Distributed Optimization by Ant Colonies. In: Varela, F., Bourgine, P. (eds.) Proceedings of the First European Conference on Artifical Life, pp. 134–142. MIT Press, Cambridge (1992)

[3] Dorigo, M., Maniezzo, V., Colorni, A.: The Ant System: Optimization by a Colony of Cooperating Agents. IEEE Transactions on Systems, Man, and Cybernetics 26, 29–41 (1996)

[4] Kennedy, J., Eberhart, R.: Particle Swarm Optimization. In: Proceedings of IEEE International Conference on Neural Networks, vol. 4, pp. 1942–1948 (1995)

[5] del Valle, Y., Venayagamoorthy, G.K., Mohaghenghi, S., Hernandez, J.C., Harley, R.G.: Particle Swarm Optimization: Basic Concepts, Variants and Applications in Power Systems. IEEE Transactions on Evolutionary Computation 12, 171–195 (2008)

[6] Seeley, T.D.: The Wisdom of the Hive. Harward University Press (1996)

[7] Teodorovic, D., Dell'orco, M.: Bee Colony Optimization-A Cooperative Learning Approach to Complex Transportation Problems, Advanced OR and AI Methods in Transportation, pp. 51–60 (2005)

[8] Passino, K.M.: Distributed Optimization and Control Using Only a Germ of Intelligence. In: Proceedings of the 2000 IEEE International Symposium on Intelligent Control, pp. 5–13 (2000)

[9] Passino, K.M.: Biomimicry of Bacteria Foraging for Distributed Optimization and Control. IEEE Control Systems Magazine 22, 52–67 (2002)

[10] Venayagamoorthy, G.K., Harley, R.G.: Swarm Intelligence for Transmission System Control. In: IEEE Power Engineering Society General Meeting, pp. 1–4 (2007)

[11] Kennedy, J., Eberhart, R.C.: Swarm Intelligence. Morgan Kaufmann Publisher, San Francisco (2001)

[12] Bai, H., Zhao, B.: A Survey on Application of Swarm Intelligence Computation to Electric Power System. In: Proceedings of the 6th World Congress on Intelligent Control and Automation, vol. 2, pp. 7587–7591 (2006)

[13] Bonabeau, E., Dorigo, M., Theraulaz, G.: Inspiration for Optimization from Social Insect Behavior. Nature 406, 39–42 (2002)

[14] Millonas, M.: Swarms, Phase Transitions, and Collective Intelligence. In: Langton, C.G. (ed.) Artificial Life III, vol. XVII, pp. 417–445. Addison-Wesley Publishing Company, Reading (1994)

[15] Dorigo, M., Gambardella, L.M.: Ant Colonies for the Traveling Salesman Problem. ByoSystems 43, 73–81 (1997)

[16] Stutzle, T., Hoos, H.: The MAX–MIN Ant System and Local Search for the Traveling Salesman Problem. In: Angeline, P. (ed.) Proceedings of the IEEE International Conference on Evolutionary Computation, pp. 308–313. Springer, Heidelberg (1997)

[17] Clerc, M., Kennedy, J.: The Particle Swarm: Explosion, Stability, and Convergence in a Multi-dimensional Complex Space. IEEE Transactions on Evolutionary Computation 6, 58–73 (2002)

[18] Mendes, R., Kennedy, J., Neves, J.: The Fully Informed Particle Swarm: Simpler, may be Better. IEEE Transactions on Evolutionary Computation 8, 204–210 (2004)

[19] von Frisch, K.: Decoding the Language of the Bee. Science 185, 663–668 (1974)

[20] Subbotina, S.A., Oleinik, A.A.: Multiagent Optimizaiton based on the Bee-Colony Method. Cybernetics and Systems Analysis 45, 177–186 (2009)

[21] Yang, X.S.: Engineering Optimizations via Nature-Inspired Virtual Bee Algorithms. In: Mira, J., Álvarez, J.R. (eds.) IWINAC 2005. LNCS, vol. 3562, pp. 317–323. Springer, Heidelberg (2005)

[22] Das, S., Abraham, A., Konar, A.: Swarm Intelligence Algorithms in Bioinformatics. In: Kelemen, A., Abraham, A., Chen, Y. (eds.) Computational Intelligence in Bioinformatics, pp. 113–147. Springer, Heidelberg (2008)

[23] Lucic, P., Teodorovic, D.: Computing with Bees: Attacking Complex Transportation Engineering Problems. International Journal on Artificial Intelligence Tools 12, 375–394 (2003)

[24] Teodorovic, D., Lucic, P., Markovic, G., Dell' Orco, M.: Bee Colony Optimization: Principles andApplications. In: Proceedings of the 8th Seminar on Neural Network Applications in Electrical Engineering, pp. 151–156 (2006)

[25] Wong, L.P., Puan, C.Y., Low, M.Y.H., Chong, C.S.: Bee Colony Optimization Algorithm with Big Valley Landscape Exploitation for Job Shop Scheduling Problems. In: Proceedings of the 40th Conference on Winder Simulation, pp. 2050–2058 (2008)

[26] Baykasoglu, A., Özbakır, L., Tapkan, P.: Artificial Bee Colony Algorithm and Its Application to Generalized Assignment Problem. Intelligence. In: Chan, F.T.S., Tiwari, M.K. (eds.) Swarm Intelligence: Focus on Ant and Particle Swarm Optimization, pp. 532–564. Itech Education and Publishing, Vienna (2007)

[27] Duan, H., Yu, X.: Progresses and Challenges of Ant Colony Optimization-Based Evolvable Hardware. In: Proceedings of the 2007 IEEE Workshop on Evolvable and Adaptive Hardware, pp. 67–71 (2007)

[28] Johnson, C., Venayagamoorthy, G.K., Palangpour, P.: Hardware Implementations of Swarming Intelligence–A Survey. In: Proceedings of the 2008 IEEE Swarm Intelligence Symposium, pp. 1–9 (2008)

2

A Review of Particle Swarm Optimization Methods Used for Multimodal Optimization

Julio Barrera and Carlos A. Coello Coello*

CINVESTAV-IPN (Evolutionary Computation Group)
Departamento de Computación
Av. IPN No. 2508, Col. San Pedro Zacatenco
México, D.F. 07360, Mexico
julio.barrera@gmail.com,ccoello@cs.cinvestav.mx

Abstract. Particle swarm optimization (PSO) is a metaheuristic inspired on the flight of a flock of birds seeking food, which has been widely used for a variety of optimization tasks [1,2]. However, its use in multimodal optimization (i.e., single-objective optimization problems having multiple optima) has been relatively scarce.

In this chapter, we will review the most representative PSO-based approaches that have been proposed to deal with multimodal optimization problems. Such approaches include the simple introduction of powerful mutation operators, schemes to maintain diversity that were originally introduced in the genetic algorithms literature (e.g., niching [3,4]), the exploitation of local topologies, the use of species, and clustering, among others.

Our review also includes hybrid methods in which PSO is combined with another approach to deal with multimodal optimization problems. Additionally, we also present a study in which the performance of different PSO-based approaches is assessed in several multimodal optimization problems. Finally, a case study consisting on the search of solutions for systems of nonlinear equations is also provided.

1 Introduction

Particle swarm optimization (PSO) is a bio-inspired metaheuristic that was proposed by James Kennedy and Russell Eberhart in 1995 [5]. PSO performs a population-based search, using particles to represent potential solutions within the search space. Each particle is characterized by its position, velocity, and a record of its past performance. Particles are influenced by their leaders, which are the best performers either from the entire swarm or their neighborhood. At each flight cycle, the objective function is evaluated for each particle, with respect to its current position, and that information is used to measure the quality of the particle and to determine the leader in the sub-swarms and the entire population.

* The second author is also affiliated to the UMI-LAFMIA 3175 CNRS.

C.P. Lim et al. (Eds.): Innovations in Swarm Intelligence, SCI 248, pp. 9–37.
springerlink.com

Although, usually, the development of optimization algorithms considers only the search of a single optimum of a given function, this is not always the case. It is possible that the function to be optimized has multiple global optima or one global optimum with many local optima in the search space. Such functions are called *multimodal* and have been widely studied in the genetic algorithms literature [3,4,6].

The PSO algorithm is a relatively recent optimization algorithm, which is quite simple, since it only consists of two rules for obtaining a new solution from a previous one. In spite of its simplicity, PSO has been found to exhibit a fast convergence to the optimum (or its vicinity) in a wide variety of optimization problems, which has significantly increased its popularity in the last few years [2]. However, until now, relatively few researchers have explored the potential of PSO for multimodal optimization, although its simplicity makes PSO a good candidate for dealing with such problems. The purpose of this chapter is precisely to review the most representative research done in this regard.

The remainder of this chapter is organized as follows. Section 2 describes the main topologies commonly adopted with PSO. Some PSO variants commonly adopted in the specialized literature are briefly described in Section 3. Section 4 introduces multimodal optimization, as well as the main approaches that have been proposed to deal with this sort of problem. Section 5 contains the test problems adopted for a small comparative study that is described and discussed in Section 6. A case study consisting of finding solutions to a system of nonlinear equations is presented in Section 7. Finally, our conclusions and some possible paths for future research are presented in Sections 8 and 9, respectively.

2 PSO Topologies

In the original PSO algorithm introduced by Kennedy and Eberhart [5] the position and velocity of a particle is updated using equations (1) and (2)

$$v_{t+1} = v_t + R_1 \cdot C_1 \cdot (g - x_t) + R_2 \cdot C_2 \cdot (p - x_t) \tag{1}$$

$$x_{t+1} = x_t + v_{t+1} \tag{2}$$

where C_1, and C_2 are the “learning” constants, R_1 and R_2 are randomly generated numbers (from a uniform distribution) in the interval $[0, 1]$, g is the position of the global best particle (i.e., the particle with the best value in the entire swarm), and p is the position with the best value recorded by the particle so far. The computation of g involves an inspection of the values of all the other particles in the swarm. In other words, any particle has access to the information of any other particle in the swarm. The global best (or *gbest*) model refers to the case in which all the particles are “connected” with each other and can transfer information among them, and it is essentially the original model of the PSO algorithm. A graphical representation of the *gbest* model is shown in Figure 1.

The *gbest* model is not the only model that has been proposed for PSO [7,8]. Another model that has been widely used is the local best (or *lbest*) model, in

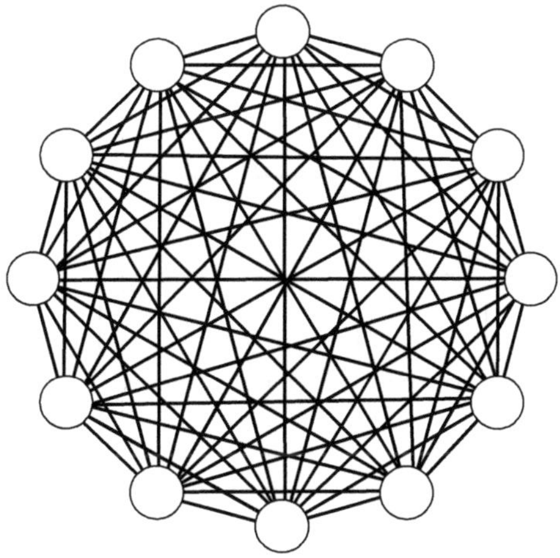

Fig. 1. Graphical representation of the *gbest* model

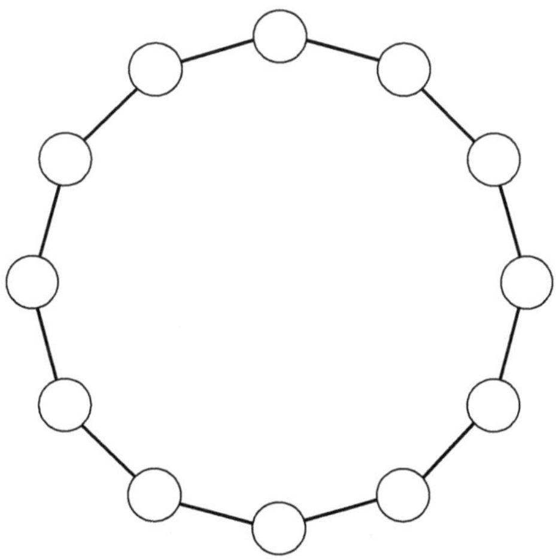

Fig. 2. Graphical representation of the *lbest* model

which a particle is connected only with k of its neighbors and can only communicate with them. The number of neighbors of each particle is usually $k = 2$. In this case, the topology of the swarm is represented as a connected ring and its graphical representation is shown in Figure 2.

Another model that has been commonly adopted in the PSO literature is the von Neumann model, in which a particle can communicate with four of its neighbors using a rectangular lattice topology. A graphical representation of the von Neumann model is shown in Figure 3. The von Neumann model is commonly adopted together with the *gbest* or *lbest* models in algorithms that use sub-swarms and a main swarm. Typically, the main swarm is arranged with

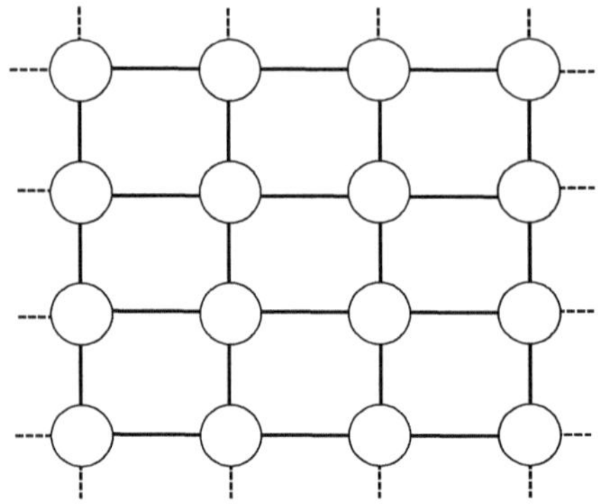

Fig. 3. Graphical representation of the von Neumann model

the von Neumann model and the sub-swarms use either the *gbest* or the *lbest* model.

3 PSO Variants

Other modifications have been added to the PSO algorithm, aiming to improve its convergence. The most common variations correspond to modifications on the computation of the velocity. Equation (1) shows the original method for computing the velocity of a particle. Eberhart and Shi [9] introduced the so-called *Inertia Weight model* in which the velocity of a particle at iteration t is multiplied by a constant parameter ω, called Inertia Weight, before computing the velocity for iteration $t+1$, as shown in equation (3).

$$v_{t+1} = \omega \cdot v_t + R_1 \cdot C_1 \cdot (g - x_t) + R_2 \cdot C_2 \cdot (p - x_t) \tag{3}$$

The parameter ω helps to balance between exploitation and exploration. Although the parameter ω is maintained constant in this model, Eberhart and Shi suggested in [10] that a linearly decreasing inertia weight may improve the convergence of the PSO algorithm. An initial ω_i and final ω_f values for the inertia are set, and the value of the inertia weight ω_t for the iteration t is computed using equation (4).

$$\omega_t = \omega_i - \frac{(\omega_f - \omega_i) \cdot t}{T} \tag{4}$$

where T is the total number of iterations and $t = 0, \ldots, T$. Another modification to the computation of the velocity is found in [11], where Clerc and Kennedy presented the so-called *Constriction Factor model.* In the Constriction Factor model not only the velocity at iteration t is multiplied by a constant, but the new computed velocity is also affected, as shown in equation (5).

$$v_{t+1} = \chi \cdot [v_t + R_1 \cdot C_1 \cdot (g - x_t) + R_2 \cdot C_2 \cdot (p - x_t)] \tag{5}$$

The constriction factor constant χ is computed using equation (6).

$$\chi = \frac{2\kappa}{|2 - \phi - \sqrt{\phi^2 - 4\phi}|} \tag{6}$$

where $\phi = C_1 + C_2$, and κ is an arbitrary constant in the range $[0, 1]$. The value of ϕ is constrained: $\phi > 4$.

The computation of the velocity of a particle involves the position of the global best particle, called *social component* and the position with the best value recorded is called *cognition component*. If one of these terms is omitted, the two resulting models are called *cognition only* and *social only*, respectively. In any of them, only one component is used for the velocity update equation, as shown in equation (7) for the cognition only model and in equation (8) for the social only model.

$$v_{t+1} = v_t + R \cdot C \cdot (p - x_t) \tag{7}$$

$$v_{t+1} = v_t + R \cdot C \cdot (g - x_t) \tag{8}$$

Both, the cognition and the social models have been used in combination with the Inertia Weight and the Constriction Factor models.

4 Multimodal Optimization Using PSO

The different models for updating the position and velocity of a particle, and the different topologies mentioned before can be considered as the basis for PSO-based multimodal approaches. In all the PSO-based multimodal approaches analyzed here, one of the three models indicated before (i.e., Inertia Weight, Decreasing Inertia Weight or Constriction Factor) is used to update the position and velocity of a particle, regardless of the particular approach adopted to deal with a multimodal problem. In methods that implement several sub-swarms and a main swarm, it is common to use two different topologies: one for the sub-swarms and another for the main swarm.

Next, we will review the most representative approaches that have been adopted to extend PSO so that it can deal with multimodal optimization problems.

4.1 Use of Mutation

Probably the easiest approach to adapt PSO to deal with multimodal optimization problems is to add a mutation operator, such as those adopted with genetic algorithms. Esquivel and Coello Coello [12] studied the use of nonuniform mutation in PSO in the context of multimodal optimization. The nonuniform mutation operator that they adopted was originally proposed in [13] for real-coded genetic algorithms, and it operates as follows. If we have the chromosome of an individual represented as a vector of real numbers $C_t = (c_1, c_2, \ldots, c_n)$ at the iteration t and c_k is the gene to mutate, then the mutated value c'_k is computed as follows:

$$c'_k = \begin{cases} c_k + \Delta(t, UB - c_k) & \text{if } P < 0.5 \\ c_k - \Delta(t, c_k - LB) & \text{otherwise} \end{cases} \tag{9}$$

where P is a randomly generated number in the interval $[0, 1]$ with a uniform distribution, LB and UB are the lower and upper bounds for the coordinate c_k, respectively, and the function $\Delta(t, z)$ returns a value in the range $[0, z]$. The probability that $\Delta(t, z)$ returns a value close to zero must increase as t increases. The function proposed in [13] for that sake is:

$$\Delta(t, z) = z \cdot \left(1 - R^{(1-\frac{t}{T})^b}\right) \tag{10}$$

where R is a randomly generated number in the range $[0, 1]$ using a uniform distribution, T is the total number of iterations, and b is a user-defined parameter that determines the degree of dependency in the total number of iterations (in [13], a value $b = 5$ is suggested). The algorithm proposed by Esquivel and Coello Coello uses in its underlying PSO, the Inertia Weight model and mutates a coordinate of the position of a particle based on the index value of the current iteration and a mutation probability P_m. The algorithm using the *gbest* model is outlined in Figure 4. A simple modification of the algorithm allows it to be used with the *lbest* model as well.

The mutation operator introduced in the PSO algorithm tries to prevent that the particles remain trapped in a local minimum.

4.2 Niching in PSO

Niching is a method originally developed for genetic algorithms which is designed to block convergence of the entire population towards a single solution (otherwise, the entire population of an evolutionary algorithm eventually converges to a single solution because of stochastic noise [14]). Niching is one of the earliest methods developed to deal with multimodality using evolutionary algorithms [3,4].

A variety of niching algorithms exist (see for example [15]). However, relatively few researchers have proposed PSO-based niching approaches. The most representative are briefly discussed next.

NichePSO. The NichePSO algorithm was proposed in [16]. This approach uses the Guaranteed Convergence Particle Swarm Optimizer from van den Bergh and Engelbrecht [17] together with the *gbest* model, aiming to prevent premature

```
swarm initialization;
for i=1 to number of particles do
    for j=1 to number of dimensions do
        Initialize x_ij with a rnd(x_max, x_min) value;
        Initialize v_ij with zero value;
        copy x_ij to p_ij;
    end
end
search the best global leader and record its position in g;
swarm flight through the search space;
repeat
    for i=1 to number of particles do
        for j=1 to number of dimensions do
            Update v_ij using p_ij and x_ij;
            Prevent explosion of v_ij;
            Update x_ij;
            if loop_number < T · P_m then
                Mutate x_ij;
            end
        end
        Evaluate fitness(x_i);
        if fitness(p_i) < fitness(x_i) then
            Update p_i;
        end
    end
until loop_number < T ;
```

Fig. 4. PSO with a nonuniform mutation operator using the *gbest* model

convergence of the particles. Additionally, the *cognition only* model is used to encourage the local search of the particles. The method used to identify niches in the NichePSO algorithm is based on keeping track of the changes in the fitness value of the particles. If a particle does not show any change within a certain number of iterations, a niche is created, containing the particle and its closest topological neighbor. The Euclidean distance is used to determine the closeness of the particles, so that the closest topological neighbor to the particle selected is the particle within the minimum distance to it. A niche radius is computed for each sub-swarm. This radius is the maximum distance between the particles in the sub-swarm and is described in equation (11).

$$r_j = \max\{||S_{x_{j,g}} - S_{x_{j,i}}||\} \tag{11}$$

This radius is used for adding particles to a sub-swarm and to merge two sub-swarms. If a particle has a distance less than r_j to the initial particle in a sub-swarm, then the particle is absorbed in the sub-swarm. If the distance between the initial particles of two sub-swarm is less than the sum of its radius, then the sub-swarms are merged. This condition is described in equation (12).

$$||S_{x_{j1,g}} - S_{x_{j2,g}}|| < (r_{j1} + r_{j2}) \tag{12}$$

If a sub-swarm has converged to an optimum it is possible that its radius r_j has a value of zero. In this case, it is difficult to determine the merge of two sub-swarms. A threshold value μ is given and if the distance between initial particles of sub-swarms with radius close to zero is less than μ, then the sub-swarms are merged. This is expressed in equation (13). The algorithm of NichePSO is outlined in Figure 5.

$$||S_{x_{j1,g}} - S_{x_{j2,g}}|| < \mu \tag{13}$$

Initialize main particle swarm;
Train main swarm particles using the cognition only model;
Update fitness of each main swarm particle;
foreach *sub-swarm* **do**
    Train sub-swarm particles using one iteration of the GCPSO algorithm;
    Update each particle's fitness;
    Update swarm radius;
end
If possible, merge sub-swarms;
Allow sub-swarms to absorb any particles from the main swarm that moved into it;
Search main swarm for any particle that meets the partition criteria;
If any is found, then create a new sub-swarm with this particle and its closest neighbor;

Fig. 5. The nichePSO algorithm

The NichePSO algorithm does not require knowing ahead of time the number of optima of the function. Also, it does not require setting up a fixed radius for the niches. However, it depends on a minimum threshold to determine when a particle does not show changes and a new niche can be created. It also depends on a minimum distance to determine when two sub-swarms that are close to converging to an optimum can be merged.

Use of sub-swarms. Another niching method is the creation of a fixed number of sub-swarms in the same search space and prevent the exploration of a same area by two or more sub-swarms. An example of this sort of method is found in the work of Zhang et al. [18] in which the Hill-Valley function [19] is used to determine if a particle belongs to a niche. If a particle does not belong to a niche, it is penalized. This prevents the exploration of more than one sub-swarm in the same area. This simple penalty rule is expressed in equation (14).

$$eval(x_i) = \begin{cases} f(x_i) & \text{if } \textit{hill-valley}(x_i, x_{best}) = 1 \\ f(x_i) - p(x_i) & \text{otherwise} \end{cases} \tag{14}$$

where $p(x)$ is a penalty function [20]. The penalty function can be static (i.e., the same penalty value is always used) with a large value for the penalty factor.

The Hill-Valley algorithm tries to determine if two points are in the same valley of a function. A set of interior points are computed following a line between the points that are being tested. The best fitness value is set as the optimal fitness for comparison purposes, with respect to the fitness values of the interior points. In case of searching for a minimum, the lowest fitness value of the points being tested is set as the minimal for comparison purposes. If all the interior points have a fitness value lower than the minimal value, then the points being tested are in the same niche. The Hill-Valley algorithm is shown in Figure 6.

```
minfit = min(fitness(i_p), fitness(i_q));
for j=0 to samples length do
    Calculate point i_interior on the line between the points i_q and i_p;
    if minfit > fitness(i_interior) then
        return 1;
    end
end
return 0;
```

Fig. 6. The Hill-Valley algorithm

The niching algorithm of Zhang et al. [18] consists of sequentially creating a given number N of sub-swarms. Each sub-swarm explores the search space until a certain criterion is fulfilled. In this case, the sub-swarm explores for a certain number of iterations. The algorithm of sequential niching is shown in Figure 7.

In the sequential PSO niching algorithm of Zhang et al. [18] is necessary to set the number of sub-swarms to be used. If there is no prior knowledge of the number of optima of the function, it is necessary to experiment with the number of niches in the algorithm. It is also required to set the number of interior points for the Hill-Valley algorithm, but unlike NichePSO, this value does not depend on threshold parameters.

The problem of choosing the number of niches and how to determine if a particle belongs to a niche is addressed by Bird and Li [21]. In that work the number of niches and their radii is adaptively computed.

To initialize the radius value, Bird and Li [21] first compute the distance between each possible pair of particles and set the initial radius to the average of the minimum distances between particles as shown in equation (15).

$$r = \frac{\sum_{i=1}^{n} d_i}{n} \tag{15}$$

where

$$d_i = min\{||p_i - p_j|| \; |j \neq i\} \tag{16}$$

Then, to create a niche, each particle is tested versus all the rest of the particles in the swarm. This information is stored in an array s that keeps track of how many steps the particles are close to each other. If this distance is less than the initial radius value for at least two particles' steps, then a niche is created.

repeat
  Run a new sub-swarm;
  $N = N + 1$;
  for $k = 1$ **to** $N - 1$ **do**
    if $hill\_valley(x_i, x_k) == 0$ **then**
      Change the fitness of x_i;
    else
      Keep the original fitness;
    end
  end
  Train the sub-swarm until a convergence criterion is met;
until *N is greater than a given value or we reached a maximum number of iterations* ;

Fig. 7. The ASNPSO algorithm

The algorithm for the creation of niches is divided in two parts: first, a graph is built, representing the particles that are close to each other within a distance less than the radius, for more than two particles' steps. The algorithm for creating the graph is shown in Figure 8.

Upon creating the graph, each particle is verified and it is grouped in a niche with the particles having a distance less than the initial radius. All particles that already belong to a niche are marked as visited in order to optimize the algorithm. The algorithm to create the niches is shown in Figure 9.

The particles that belong to a niche are updated using the Constriction Factor equations, adopting the *gbest* model in each niche. The particles that do not belong to a niche form a main swarm that uses a von Neumann topology and are updated using the Constriction Factor model. At each iteration of the algorithm, the niches are recalculated, which affects the performance of the algorithm.

```
Determine r using equation (15);
Create an undirected graph G containing a node for each particle, but no edges;
for i = 1 to n − 1 do
    for j = i + 1 to n do
        if ||p_i − p_j|| < r then
            Increment s_ij;
            if s_ij < 4 then
                s_ij ← 4;
            end
            if s_ij >= 2 then
                Create an edge in G from p_i to p_j;
            end
        else
            Decrement s_ij;
            if s_ij < 0 then
                s_ij ← 0;
            end
        end
    end
end
```

Fig. 8. Procedure to create the niches graph

```
Create a set variable visited ← ∅;
Create a set variable reachable;
for i=1 to n do
    if p_i ∉ visited and d_i < r then
        reachable ← {p_i, p_j ∈ P|p_j ∀j reachable from p_i ∈ G};
        Create a new niche s ← {p_j};
        for p ∈ reachable and p ∉ visited do
            visited ← visited ∪ {p};
            s ← s ∪ {p};
        end
    end
end
```

Fig. 9. Creating the niches from the graph G

4.3 Clustering Techniques in PSO

Clustering into a PSO algorithm was first introduced by Kennedy [22]. The method of "stereotyping" adopted by Kennedy, consists of using the k-means algorithm to partition the main swarm into k sub-swarms. The k-means algorithm is shown in Figure 10. In each sub-swarm, the position of the global best is replaced by the position of the centroid of the sub-swarm.

```
Initialize k centroids;
repeat
    Compute the distance of each point to the k clusters;
    Assign each point to the nearest cluster;
    Recompute the centroid of each cluster;
until stop criteria is met ;
```

Fig. 10. k-means algorithm

In his work, Kennedy uses the Constriction Factor model for the PSO algorithm and explores the effects of exchanging the global best g, and the best recorded position p of a particle by the centroid of the sub-swarm in both the *gbest* and the *lbest* models.

A more recent work by Passaro and Starita [23] tries to improve the previous clustering approach by introducing several modifications. First, the number k of sub-swarms is estimated by testing different values of k and by using a Bayesian information criterion to decide which value of k is optimal [24]. The authors also limit the number of particles in each sub-swarm to correspond to the mean of the number of particles in each sub-swarm after the initial partition. If a cluster exceeds the mean of the number of particles, the particles with the worst fitness values are removed and added to the main swarm. The particles in the main swarm are updated separately using the *lbest* model with a von Neumann topology. The algorithm for identifying a niche is shown in Figure 11.

```
Cluster particle's pbest with the k-means algorithm;
Compute the average number of particles per cluster, N_avg;
Set N_u = 0;
foreach Cluster C_j do
    if N_j > N_avg then
        remove the N_j - N_avg particles from C_j;
        add N_j - N_avg to N_u;
    end
    Adapt the neighborhood structure for the particles in C_j;
end
Reinitialize the N_u un-niched particles;
```

Fig. 11. Algorithm for the identification of niches

The algorithm of Passaro and Starita also recalculates the niches at intervals of c iterations. The procedure for indentifying niches is used to form again the sub-swarms and the main swarm with the *lbest* model. The algorithm is shown in Figure 12.

```
Initialize particles with random positions and velocities;
Set particles' pbest to their current positions;
Calculate particles' fitness;
for t = 0 to T − 1 do
    if t mod c = 0 then
        Execute the procedure Identify Niches;
    end
    Update particles' velocities;
    Update particles' positions;
    Recalculate particles' fitness;
    Update particles' and neighborhood best positions;
end
```

Fig. 12. The *k*PSO algorithm

Although the optimal number of niches is computed, it is necessary to setup a maximum value of niches to carry out this calculation. It is also needed to determine the number of iterations c before recalculating the niches.

4.4 PSO with Species

The use of species for dealing with multimodal optimization problems, within a genetic algorithms context, is introduced in Li et al. [25]. In this method, individuals with high fitness values are selected as "seeds" to form clusters of individuals called "species", around the seeds. The procedure to select the seeds and form species consists of the following steps:

1. The best individual of a population is selected as a seed.
2. All the individuals with a distance less than a parameter value r from the seed are considered to belong to the same species.
3. The above procedure is repeated by selecting the next seed from the particles that do not belong to a species until all the particles are part of one.

The seeds selected in a generation are reintroduced in the next generation by comparing the fitness values of the new individuals within the species. If the individual with the worst fitness value in the species has a worse fitness value than the seed, then the seed replaces the worst individual of the species. Otherwise, the seed replaces the worst individual in the population even if the seed has a worse fitness value than the individual being replaced.

The concept of species was introduced into PSO by Li [26], with some changes. First, the species' seeds are not conserved nor reintroduced into the swarm. Also, the position of the seed of a species replaces the position of the global best for each particle in the species. The algorithm for the creation of the species is shown in Figure 13.

After the creation of the species, all particles are updated using the Constriction Factor equations. The algorithm for the species-based PSO approach is shown in Figure 14.

```
S = ∅;
while the end of L_sort has not been reached do
    found ← FALSE;
    forall p ∈ S do
        if d(s,p) ≤ R_s then
            found ← TRUE;
            break;
        end
    end
    if Not found then
        S ← S ∪ {s}
    end
end
```

Fig. 13. The algorithm for determining the species' seeds

```
Generate the initial population;
repeat
    Evaluate all particles in the population;
    Sort particles in descending order of their fitness value;
    Determine the species' seeds for the current population;
    Assign each species' seed identified as the g particle to all individuals
    identified in the same species;
    Update the velocity and position of all the particles;
until Termination condition ;
```

Fig. 14. The species-based PSO

In the Species Particle Swarm Optimization algorithm, it is not necessary to set the number of sub-swarms that will be created before starting to iterate, since such number depends on the radius parameter, which is a value that needs to be set by the user.

4.5 Other Methods for Multimodal Optimization

The niching and species methods are not the only approaches that have been used with the PSO algorithm, since several other techniques have been proposed in the specialized literature as well. For example, a cooperative PSO was presented by van den Bergh and Engelbrecht [27]. This approach follows the idea from Potter and de Jong [28] who proposed that the population of an evolutionary algorithm is divided into sub-populations that cooperate unlike the niche methods in which the sub-populations compete. Initially, in the Cooperative PSO algorithm, the main swarm is divided into several sub-swarms, each of which searches in only one dimension of decision variable space. Each sub-swarm cooperates by searching one coordinate of the solution. To evaluate the fitness function of a particle in a sub-swarm, an auxiliary vector is used. This vector has as its coordinates to the position of the particle with the best fitness value of each sub-swarm. In order to evaluate a particle in the jth sub-swarm the jth coordinate of the auxiliary vector is replaced with the position of the particle, and the fitness value of the particle is set to the fitness of the auxiliary vector. If some of the variables of the fitness function are correlated, it is preferred that the search space of a sub-swarm includes the correlated variables. This can be done easily with the partition model of the Cooperative PSO, but the information of correlated variables is not always available. An approach to try to group correlated variables is to partition the search space arbitrarily into k subspaces, with $k < D$ and D the dimension of the search space, as to create k sub-swarms that search into each sub-space, separately. The Cooperative PSO algorithm has a drawback: it can be trapped in pseudo-minima [27]. In order to overcome this problem, a phase is added to the Cooperative PSO algorithm, in which all the sub-swarms are considered as a single main swarm and the Constriction Factor model is used to update the position and velocity of the particles. This is called the Hybrid Cooperative PSO and its algorithm is shown in Figure 15.

The number of sub-swarms and the dimension of the subspace in which they search are randomly set and may not be the best possible for the fitness function being optimized.

Other method that implements in part a strategy similar to the Cooperative PSO is the Comprehensive Learning PSO developed by Liang et al. [29]. This algorithm uses a *social only* model in which only the social component of the velocity update equation is used together with the Inertia Weight model. Thus, to compute the velocity of a particle, equation (17) is used.

$$v_{t+1,d} = \omega \cdot v_{t,d} + C \cdot R_d \cdot (p_{g,f(d)} - x_d) \tag{17}$$

where the subindex d corresponds to the dimension, and $p_{g,f(d)}$ is an exemplar computed for the particle. Exemplars are computed using an idea similar to the one introduced by van den Bergh and Engelbrecht [27] and replace the position of the global best particle. To compute an exemplar, a coordinate of the $p_{g,f(d)}$ position is selected according to the following method:

1. A random number $rand$ is computed for each coordinate d. If $rand >= Pc_i$ (Pc_i is a probability for the particle i), then the value of the dth coordinate of the best recorded position of the particle i is copied to the dth coordinate of the exemplar.
2. Otherwise, if $rand < Pc_i$, then two particles are randomly selected from the swarm, so that they are different from particle i. Additionally, a tournament-like selection is done with the two particles selected. Also the dth coordinate of the best recorded position of the particle with a better fitness value is copied to the dth coordinate of the exemplar.
3. If the condition $rand < Pc_i$ is never true for any coordinate d, then a random number k is computed with $k \in 1, \ldots, D$ (D is the number of decision variables), and the procedure of step 2 is repeated for the coordinate k.

```
b(j, z) = (g_1, ..., g_{j-1}, z, g_{j+1}, ..., g_k);
K_1 = n mod K;
K_2 = K − (n mod K);
Initialize K_1⌈n/K⌉-dimensional PSOs: P_j, j ∈ [1..K_1];
Initialize K_2⌊n/K⌋-dimensional PSOs: P_j, j ∈ [(K_1 + 1)..K];
Initialize an n-dimensional PSO: Q;
repeat
    foreach swarm i ∈ [1..K] do
        foreach particle j ∈ [1..s] do
            if f(b(i, x_ij)) < f(b(i, p_ij)) then
                p_ij = x_ij;
            end
            if f(b(i, p_ij)) < f(b(i, g_ij)) then
                g_ij = p_ij;
            end
        end
        Perform updates on P_j;
    end
    Select random k ~ U(1, s/2) such that p_k ≠ g;
    x_k = b(1, g_1);
    foreach particle j ∈ [1..s] do
        if f(x_j) < f(p_j) then
            p_j = x_j;
        end
        if f(p_j) < f(g) then
            g = p_j;
        end
    end
    Perform updates in Q;
    foreach swarm j ∈ [1..K] do
        Select random k ~ U(1, s/2) such that p_jk ≠ g_j;
        x_jk = g_j;
    end
until stopping condition is true ;
```

Fig. 15. Pseudocode for the CPSO-H_k algorithm

To compute the value of the probability Pc_i for the particle i, the authors use an empirical relation. For example, to compute the probability values Pc_i in the range $[0.05, 0.5]$, equation (18) is used.

$$Pc_i = 0.05 + 0.45 \cdot \frac{e^{\left(\frac{10(i-1)}{ps-1}\right)} - 1}{e^{10} - 1} \tag{18}$$

where *ps* is the number of particles in the swarm. The procedure for computing an exemplar is shown in Figure 16.

```
for d=1 to D do
    if rand < Pc_i then
        f1_d = ⌈rand1_d · ps⌉;
        f2_d = ⌈rand2_d · ps⌉;
        if fitness(p_{f1_d}) > fitness(p_{f2_d}) then
            f_d = f_{1d};
        else
            f_d = f_{2d};
        end
    else
        f_d = i;
    end
end
```

Fig. 16. Selection of the exemplar dimension for particle i

The Comprehensive Learning PSO algorithm also keeps track of the changes in the fitness of the particles. If the particle does not show changes in its best recorded position after m iterations, then a new exemplar is computed for the particle. The value for m is arbitrary and the authors use $m = 7$, which was empirically obtained. The algorithm also uses a policy of not updating the fitness value nor the best recorded position if the particle is out of the allowable search space. It also updates the global best of the swarm each time that a particle is updated. This is similar to the evaluation in the case of a *lbest* topology. The algorithm of the Comprehensive Learning PSO is shown in Figure 17.

The Comprehensive Learning PSO algorithm requires several parameters: the probability Pc_i of computing an exemplar, the refreshing gap m to re-compute an exemplar if the best position of a particle does not present changes (this value is based on the Decreasing Inertia PSO). Additionally, an initial ω_i and a final ω_f must also be set by the user. In spite of the number of parameters to set and its relative complexity, Comprehensive Learning PSO has obtained better results than many other PSO-based methods in a variety of multimodal optimization problems.

4.6 Hybrid Methods

The PSO algorithm has also been hybridized with other algorithms in an attempt to improve its performance in multimodal optimization problems. That is the case of the hybrid of PSO and Differential Evolution (DE) introduced by Pant et al. [30]. This algorithm works in two phases. First, at each iteration, a mutant of each particle is computed by adopting the same procedure from the DE algorithm. Then, the crossover operator from DE is applied. If the new particle created with the DE procedure has a better fitness value than the particle being compared, then the position of the particle is replaced by the position of the newly generated particle. If the fitness of the new particle is worse, then the particle is removed and a new particle is generated. The fitness value of the particle with the updated position is compared with the fitness value of the particle in the previous position. If the new particle has a better fitness value, then the position of the particle is updated to the new computed position using the equations of the PSO algorithm. If the fitness value is worse, then the position of the particle remains unchanged. The algorithm of the hybrid of DE and PSO is shown in Figure 18.

```
Initialize swarm;
for k = 1 to max_gen do
    ω = ω0 − (ω0 − ω1) · k/max_gen;
    for i = 1 to ps do
        if flag_i >= m then
            Select an exemplar for particle i;
            flag_i = 0;
        end
        Update velocity and position of the particles;
        if x_i ∈ [X_min, X_max] then
            Update fitness x_i;
            if fitness(x_i) > fitness(p_i) then
                p_i = x_i;
                flag_i = 0;
                if fitness(x_i) > fitness(g) then
                    g = x_i;
                end
            else
                flag_i = flag_i + 1;
            end
        end
    end
end
```

Fig. 17. The CLPSO algorithm

A similar hybrid is presented in the work of Shelokar et al. [31], in which a hybrid of PSO and ant colony optimization (ACO) is presented. In this hybrid, a colony of ants is created within the swarm of particles. Each ant is related to a particle, and after a particle is updated, its related ant is also updated, by following a simple pheromone scheme. If the fitness value of the ant is better than the fitness value of the particle, then the position of the particle is replaced with the position of the ant. The algorithm of this hybrid is shown in Figure 19 where Z_t^i is computed according to equation (19).

$$z_t^i = N(g, \sigma) \tag{19}$$

```
Initialize swarm;
for i = 1 to N do
    Select r1, r2, r3 ∈ N;
    for j = 1 to D do
        Select j_rand ∈ D;
        if rand() < Cr or j = j_rand then
            U_{ji,g+1} = x_{r1,g} + F · (x_{r2,g} − x_{r3,g});
        end
        if f(U_{ji,g+1}) < f(x_{ji,g}) then
            x_{ji,g+1} = U_{ji,g+1};
        else
            Compute tx_{ji} using PSO equations;
            if f(tx_{ji}) < f(x_{ji,g}) then
                x_{ji,g+1} = tx_{ji};
            else
                x_{ji,g+1} = x_{ji,g};
            end
        end
    end
end
```

Fig. 18. Hybrid of PSO and DE

```
Initialize swarm;
Initialize ants;
repeat
    Update particles' position and velocity;
    Evaluate the fitness function of each particle;
    Generate P solutions z_t^i using equation (19);
    for i = 1 to ps do
        if f(z_t^i) < f(x_i) then
            x_i = z_t^i;
        end
        Update p_i of particles;
    end
    Update g best;
until t = T ;
```

Fig. 19. Hybrid of PSO and ACO

where $N(g, \sigma)$ is a normal distribution with mean g and standard deviation σ. The value of σ is decreased at each iteration by multiplying it by a constant d with values in the range $[0.25, 997]$. If the value of σ is less than a lower bound σ_m, then the value of σ is set to $\sigma = \sigma_m$ and remains constant for the rest of the iterations.

5 Test Functions

In order to have a better idea of the differences in performance of some of the approaches previously discussed, we decided to perform a small comparative study. For that sake, we used a set of five test functions commonly adopted in the multimodal optimization literature. The selected problems are: Rastrigin's function [32,33], Griewank's function [34], Ackley's function [35,36], Schwefel's function [37], and Shubert's function. Equations (20) to (24) show the test functions in the same order as indicated above.

$$f_1(x) = nA + \sum_{i=1}^{n} x_i^2 - A\cos(\omega x_i) \tag{20}$$

$$f_2(x) = \frac{1}{4000}\sum_{i=1}^{n} x_i^2 - \prod_{i=1}^{n} \cos\left(\frac{x_i}{\sqrt{i}}\right) \tag{21}$$

$$f_3(x) = -20e^{-0.2\sqrt{\frac{1}{30}\sum_{i=1}^{n} x_i^2}} - e^{\frac{1}{30}\sum_{i=1}^{n}\cos(2\pi x_i)} + 20 + e \tag{22}$$

$$f_4(x) = 418.9829n + \sum_{i=1}^{n} x_i \sin\left(\sqrt{|x_i|}\right) \tag{23}$$

$$f_5(x) = \prod_{i=1}^{n}\sum_{j=1}^{5} j\cos\left[(j+1)x_i + j\right] \tag{24}$$

Except for Shubert's test problem, which has multiple global optima, the rest of the test functions adopted have only one global optimum, but contain several local optima. Shubert's test problem has several global optima distributed uniformly throughout the search space. In Table 1, we show a summary of the features of the test functions adopted, including their search range, the number of optima in the search range and the position of their global optima.

Table 1. Summary of the features of the test functions adopted

Test Function	Optimum	Optimum fitness value	Search range
Rastrigin	$[0, \ldots, 0]$	0	$[-5.12, -5.12]^D$
Griewank	$[0, \ldots, 0]$	0	$[-600.0, 600.0]^D$
Ackley	$[0, \ldots, 0]$	0	$[-15.0, 30.0]^D$
Schwefel	$[1, \ldots, 1]$	0	$[-500.0, 500.0]^D$
Shubert	18 (in two dimensions)	-186.7309	$[-10.0, 10.0]^2$

6 Experiments and Results

The algorithms applied to the test functions are the following: PSO with mutation [12], species PSO [25], comprehensive learning PSO [29], and the hybrid of PSO and ACO [31].

Table 2. Mean of the best fitness value after 5000 iterations.

	CLPSO	APSO	SPSO	MPSO
Rastrigin	1.5	188.56	175.44	436.33
Griewank	1.68	838.41	15.75	0.97
Ackley	1.96e-07	19.36	10.71	3.59
Schwefel	13.29	1.28e+13	6.81e+9	6.21e+4
Shubert	-186.56	-185.87	-79.94	136.88

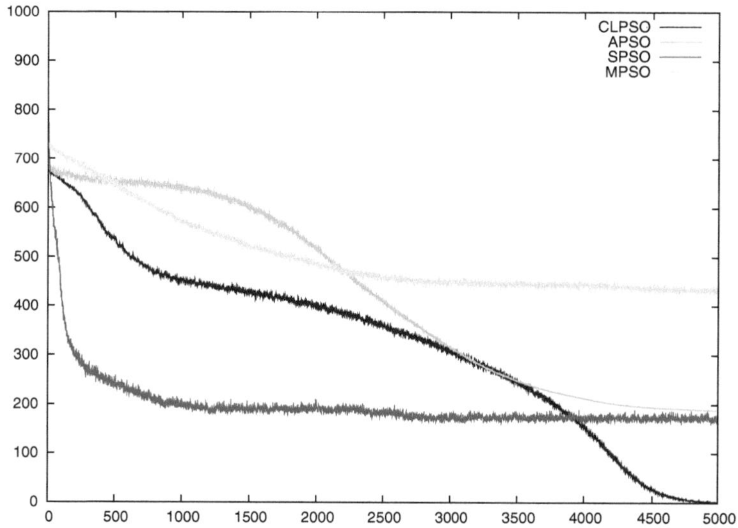

Fig. 20. Diagram of convergence for the four algorithms applied to Rastrigin's function.

Next, we describe the setup for each algorithm. It is important to note that the values adopted for the parameters of each approach follow those proposed by their authors.

For the PSO with mutation we used: swarm size = 40 particles, learning constant values $C_1 = C_2 = 1.3$, inertia weight $\omega = 0.3$, mutation probability $P_m = 0.9$.

The PSO with species used: swarm size = 40 particles, learning constants values $C_1 = C_2 = 2.05$, radius value between 1/10 and 1/20 of the allowable variables range.

For the Comprehensive Learning PSO we used: swarm size = 40 particles, initial and final inertia values $\omega_i = 0.9$, and $\omega_f = 0.4$, learning constant values $C_1 = C_2 = 1.49455$, refreshing gap $m = 7$.

Finally, for the hybrid of PSO and ACO we used: swarm size = 40 particles, learning constant values $C_1 = C_2 = 2.0$, initial and final inertia values $\omega_i = 0.7$, $\omega_f = 0.4$, $d = 0.5$, $\sigma_m = 10^{-4}$.

In order to allow a fair comparison, all the algorithms performed 5,000 iterations. The convergence plots for all the test functions are shown in Figures 20,

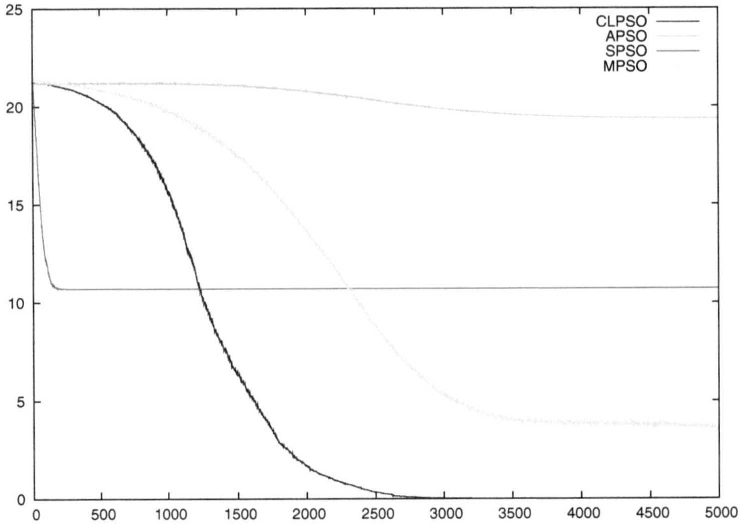

Fig. 21. Diagram of convergence for the four algorithms applied to Ackley's function.

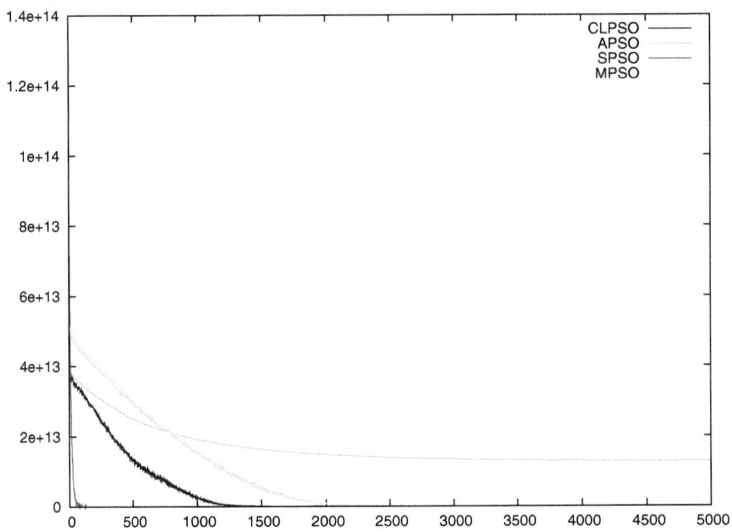

Fig. 22. Diagram of convergence for the four algorithms applied to Schwefel's function.

21, 22, 23 and 24. Table 2 shows the mean of the best fitness values obtained by the algorithms after 5000 iterations for the test functions adopted.

From the plots shown in Figures 20 to 24, and from the summary shown in Table 2, we can observe that the Comprehensive Learning PSO algorithm has the best convergence and it is consistent in all the test functions adopted. The

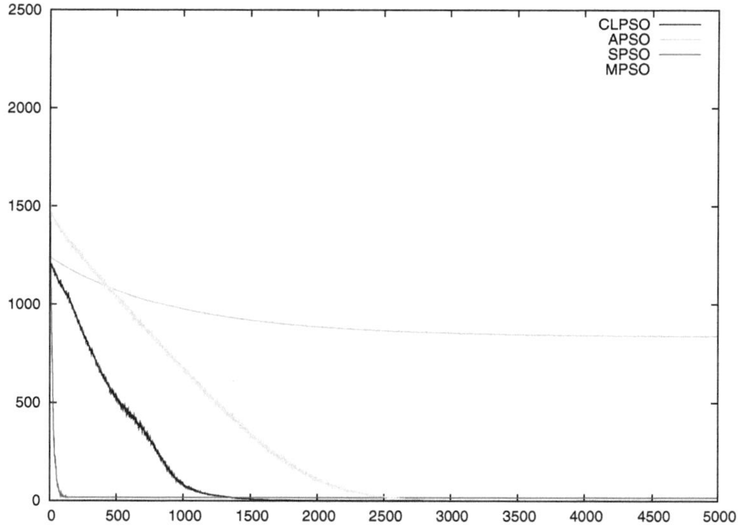

Fig. 23. Diagram of convergence for the four algorithms applied to Griewank's function.

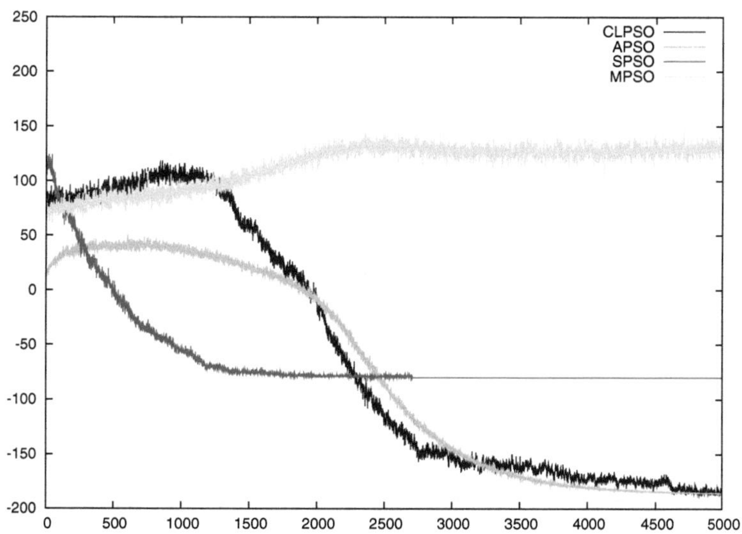

Fig. 24. Diagram of convergence for the four algorithms applied to Shubert's function.

Species PSO algorithm depends strongly on the selection of the radius, but it tends to stabilize faster than the others. The PSO with mutation has a lower rate of convergence but it shows good results in almost all the test functions tested, which seems to indicate that the simple introduction of a good mutation operator may be enough if the cost of evaluating the objective function is not significant. The hybrid of PSO and ACO had the worst results both in terms of convergence rate and accuracy, but it shows good results and better stability than any other algorithm in Shubert's test function.

In Figure 24 we show the convergence graphs for Shubert's test function. Although it is analyzed only in two dimensions, this test function has 18 global optima and, as we can observe from Figure 24, for almost all the algorithms, the value of the reported optimum shows oscillations. In the case of the PSO with mutation the reported optimum value tends to diverge.

7 Search of Solutions of a System of Nonlinear Equations

The study of dynamical systems is of great importance in almost all fields of knowledge. A dynamical system is commonly modeled using a system of ordinary differential equations such as the one represented in equation (25).

$$\dot{x} = f(x) \tag{25}$$

with $x \in R^n$. The first step in the analysis of a dynamical system is to find a fixed point of the system. A fixed point is a point in which all the derivatives are equal to zero, that is, a point x' such that

$$0 = f(x') \tag{26}$$

To find a fixed point we need to find a solution of the system of equations $f(x)$ that most of the time is nonlinear.

An example of the physical interpretation of this sort of system is the electrical power system of three nodes [38,39] shown in Figure 25.

The corresponding dynamic system is modeled with equations (27) to (30).

$$\dot{\delta}_m = \omega \tag{27}$$

$$\begin{aligned}\dot{\omega} &= \frac{1}{M}(P_m - D_m\omega + VE_mY_m\sin(\delta - \delta_m - \theta_m) + \\ &\quad E_m^2Y_m\sin\theta_m)\end{aligned} \tag{28}$$

$$\dot{\delta} = \frac{1}{K_{qw}}(-K_{qv2}V^2 - K_{qv}V + Q - Q_0 - Q_1) \tag{29}$$

$$\begin{aligned}\dot{V} &= \frac{1}{TK_{qw}K_{pv}}[-K_{qw}(P_0 + P_1 - P) + (K_{pw}K_{qv} - K_{qw}K_{pv})V + \\ &\quad K_{pw}(Q_0 + Q_1 + Q) + K_{pw}K_{qv2}V^2]\end{aligned} \tag{30}$$

where

$$\begin{aligned}P = &-VE_0'Y_0'\sin(\delta + \theta_0') - VE_mY_m\sin(\delta - \delta_m + \theta_m) \\ &+V^2(Y_0'\sin\theta_0' + Y_m\sin\theta_m)\end{aligned} \tag{31}$$

$$\begin{aligned}Q = &\,VE_0'Y_0'\cos(\delta + \theta_0') + VE_mY_m\cos(\delta - \delta_m + \theta_m) \\ &-V^2(Y_0'\cos\theta_0' + Y_m\cos\theta_m)\end{aligned} \tag{32}$$

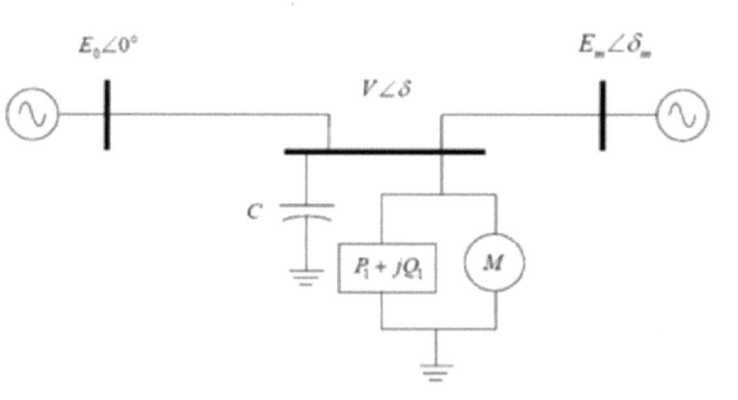

Fig. 25. Diagram of a power system of three nodes

Table 3. Constant values for the symbols in the system of nonlinear equations

symbol	value	symbol	value	symbol	value
K_{pw}	0.4	K_{pv}	0.3	K_{qw}	-0.03
K_{qv2}	2.1	T	8.5	K_{qv}	-2.8
P_0	0.6	P_1	0.0	Q_0	1.3
E_0'	2.5	P_m	1.0	E_m	1.0
M	0.3	Y_m	5.0	Y_0'	8.0
θ_0'	12.0	Q_1	10.0	θ_m	-5.0
D_m	0.05				

Table 4. Values of the search ranges for the decision variables

variable	range
δ_m	$[0, 1]$
ω	$[-1, 1]$
δ	$[0, 1]$
V	$[0, 2]$

Only four variables are considered: δ_m, ω, δ, and V. All the other symbols are set to constant values as shown in Table 3. We need to find a fixed point in the search space defined by the range values for the variables shown in Table 4.

For this example, we used the Comprehensive Learning PSO with a swarm size of 10 particles, initial and final inertia values $\omega_i = 0.9$, $\omega_f = 0.4$, learning constant values $C_1 = C_2 = 1.49455$, refreshing gap $m = 7$ and a total of $1,000$ iterations. The fitness function is $|f(x)|$ with f being the system of equations (27) to (30) and $x = (\delta_m, \omega, \delta, V)$. After applying Comprehensive Learning PSO, a solution is found in the point $(0.047, 0.0, 0.310, 1.35)$ with an error of 10^{-4} that is an acceptable error value for this application.

The PSO algorithm has several advantages when adopted for searching solutions for systems of nonlinear equations: PSO does not require of a "good" initial point to perform the search, and the search space can be bounded by lower and upper values for each decision variable. Additionally, no continuity or differentiability of the objective function is required. What can be considered as the main disadvantage of PSO in this sort of application is its relatively poor accuracy, which is caused by the coarse granularity of the search performed by the algorithm. This can, of course, be improved either by running the PSO algorithm for a larger number of iterations (although at a higher computational cost) or by post-processing the solution produced by the PSO algorithm with a traditional numerical optimization technique.

8 Conclusions

In this chapter, we have shown a review of the most representative PSO-based methods available for multimodal optimization. As we saw, most of these methods were either adapted or inspired by research reported in the genetic algorithms literature. There are, however, other methods that are not directly derived from such literature, because they rely on hybridizations between PSO and another metaheuristic (e.g., differential evolution or ant colony optimization), with the clear aim of benefitting from the advantages of both types of approaches.

A particular case is the clustering method originally developed by Kennedy [22] and further modified by Bird and Li [21], since this sort of approach was specifically designed for a PSO algorithm.

It is worth mentioning that all the methods analyzed here have been developed using as a basis the known models for updating the position and velocity of the PSO algorithm, namely the Inertia Weight and the Constriction Factor. To the authors' best knowledge, there are no models currently available for updating the position and velocity of a particle, that had been specifically developed for dealing with multimodal optimization problems. This is also true for the topologies of the swarm, since the authors are not aware of any new topologies that had been developed specifically for multimodal problems. It is worth noting, however, that in the method for computing the niche parameters for Bird and Li's algorithm [21], a graph based on the closeness of the particles is used to determine if a particle belongs to a niche. Nevertheless, once the sub-swarms are built they use a *gbest* topology in each sub-swarm and a von Neumann topology with the particles that do not belong to a sub-swarm, but are part of the main swarm. A particular case is the Comprehensive Learning PSO algorithm. Although in this algorithm the Inertia Weight is used together with the *social only* model, the position of the best particle in the swarm is replaced with a computed exemplar for each particle, which is an *ad-hoc* method specifically developed for this approach.

In general, all the methods studied show advantages over the use of a basic PSO algorithm, but they also introduce new parameters that need to be set by the user. In most cases, the parameters were empirically tuned after performing a series of experiments [29]. The current research in the area includes the development of methods for adaptively computing the parameters required for dealing with multimodal optimization problems (e.g., [21]).

We have also conducted a small comparative study in which we analyzed the performance of several of the algorithms discussed in this chapter. This study showed that Comprehensive Learning PSO had the best overall results both in terms of quality of the final solutions and of consistency in reaching such results.

Finally, we presented an application of PSO in a case study in which the problem is multimodal: the search of solutions to a set of nonlinear equations. The results in this case were very encouraging and showed some of the advantages of adopting PSO with respect to traditional numerical optimization techniques (e.g., the use of bounded decision variables, and the use of randomly generated solutions to initiate the search).

9 Future Work

There are several areas that remain open for those interested in working in this area. First, the development of new PSO-based methods for multimodal optimization still has a lot to offer. For example, the development of new approaches that have less (or no) parameters that have to be fine-tuned by the user and that remain effective in a variety of test problems is an interesting venue for future research.

There are approaches, such as those based on clustering methods, that do not use any information from the fitness values of the particles. Such information could be used, for example, to improve the computation of the centroids of the clusters.

Finally, the development of additional hybrid methods is also an interesting path for future research, that can give us more insights regarding the behavior and potential advantages and disadvantages of different metaheuristics.

References

1. Kennedy, J., Eberhart, R.C., Shi, Y.: Swarm intelligence. Morgan Kaufmann, San Francisco (2001)
2. Engelbrecht, A.P.: Fundamentals of Computational Swarm Intelligence. John Wiley & Sons, Chichester (2006)
3. Goldberg, D.E., Richardson, J.: Genetic algorithm with sharing for multimodal function optimization. In: Grefenstette, J.J. (ed.) Genetic Algorithms and Their Applications: Proceedings of the Second International Conference on Genetic Algorithms, pp. 41–49. Lawrence Erlbaum, Hillsdale (1987)
4. Deb, K., Goldberg, D.E.: An Investigation of Niche and Species Formation in Genetic Function Optimization. In: Schaffer, J.D. (ed.) Proceedings of the Third International Conference on Genetic Algorithms, San Mateo, California, George Mason University, pp. 42–50. Morgan Kaufmann Publishers, San Francisco (1989)
5. Kennedy, J., Eberhart, R.C.: Particle swarm optimization. In: Proceedings of IEEE International Conference on Neural Networks, pp. 1942–1948. IEEE Press, Piscataway (1995)

6. Deb, K., Kumar, A.: Real-coded Genetic Algorithms with Simulated Binary Crossover: Studies on Multimodal and Multiobjective Problems. Complex Systems 9, 431–454 (1995)
7. Kennedy, J.: Small worlds and mega-minds: effects of neighborhood topology on particle swarm performance. In: Proceedings of the 1999 IEEE Congress on Evolutionary Computation (CEC 1999), vol. 3, pp. 1931–1938. IEEE Computer Society Press, Los Alamitos (1938)
8. Kennedy, J., Mendes, R.: Population structure and particle swarm performance. In: Proceedings of the 2002 IEEE Congress on Evolutionary Computation (CEC 2002), Washington, DC, USA, pp. 1671–1676. IEEE Computer Society Press, Los Alamitos (2002)
9. Shi, Y., Eberhart, R.: A modified particle swarm optimizer. In: Proceedings of the 1998 IEEE International Conference on Evolutionary Computation Proceedings, pp. 69–73. IEEE Press, Los Alamitos (1998)
10. Shi, Y., Eberhart, R.C.: Empirical study of particle swarm optimization. In: Proceedings of the 1999 IEEE Congress on Evolutionary Computation (CEC 1999), vol. 3, pp. 1945–1950. IEEE Press, Los Alamitos (1999)
11. Clerc, M., Kennedy, J.: The particle swarm: explosion, stability, and convergence in a multidimensional complex space. IEEE Transactions on Evolutionary Computation 6(1), 58–73 (2002)
12. Esquivel, S.C., Coello Coello, C.A.: On the use of particle swarm optimization with multimodal functions. In: Proceedings of the 2003 IEEE Congress on Evolutionary Computation (CEC 2003), vol. 2, pp. 1130–1136. IEEE Press, Los Alamitos (2003)
13. Michalewicz, Z.: Genetic algorithms + data structures = evolution programs, 3rd edn. Springer, London (1996)
14. Goldberg, D.E.: Genetic Algorithms in Search, Optimization and Machine Learning. Addison-Wesley Publishing Co., Reading (1989)
15. Mahfoud, S.W.: Niching Methods for Genetic Algorithms. PhD thesis, University of Illinois at Urbana-Champaign, Department of General Engineering, Urbana, Illinois (May 1995)
16. Brits, R., Engelbrecht, A., van den Bergh, F.: A Niching Particle Swarm Optimizer. In: Wang, L., et al. (eds.) Proceedings of the 4th Asia-Pacific Conference on Simulated Evolution and Learning (SEAL 2002), Orchid Country Club, Singapore, vol. 2, pp. 692–696. Nanyang Technical University (2002)
17. van den Bergh, F., Engelbrecht, A.: A new locally convergent particle swarm optimiser. In: 2002 IEEE International Conference on Systems, Man and Cybernetics, vol. 3. IEEE Press, Los Alamitos (2002)
18. Zhang, J., Huang, D.S., Liu, K.H.: Multi-Sub-Swarm Particle Swarm Optimization Algorithm for Multimodal Function Optimization. In: 2007 IEEE Congress on Evolutionary Computation (CEC 2007), Singapore, pp. 3215–3220. IEEE Computer Society Press, Los Alamitos (2007)
19. Ursem, R.K.: Multinational evolutionary algorithms. In: Proceedings of the Congress on Evolutionary Computation, pp. 1633–1640. IEEE Press, Los Alamitos (1999)
20. Yeniay, Özgür: Penalty Function Methods for Constrained Optimization with Genetic Algorithms. Mathematical and Computational Applications 10(1), 45–56 (2005)
21. Bird, S., Li, X.: Adaptively Choosing Niching Parameters in a PSO. In: Tiwari, M.K., et al. (eds.) 2006 Genetic and Evolutionary Computation Conference (GECCO 2006), Seattle, Washington, USA, vol. 1, pp. 3–9. ACM Press, New York (2006)

22. Kennedy, J.: Stereotyping: improving particle swarm performance with cluster analysis. In: Proceedings of the, Congress on Evolutionary Computation, vol. 2, pp. 1507–1512. IEEE Computer Society Press, Los Alamitos (2000)
23. Passaro, A., Starita, A.: Particle swarm optimization for multimodal functions: a clustering approach. Journal of Artificial Evolution and Applications 8(2), 1–15 (2008)
24. Pelleg, D., Moore, A.: X-means: Extending k-means with efficient estimation of the number of clusters. In: Proceedings of the Seventeenth International Conference on Machine Learning, pp. 727–734. Morgan Kaufmann, San Francisco (2000)
25. Li, J.P., Balazs, M.E., Parks, G.T., Clarkson, P.J.: A species conserving genetic algorithm for multimodal function optimization. Evolutionary Computation 10(3), 207–234 (2002)
26. Li, X.: Adaptively choosing neighbourhood bests using species in a particle swarm optimizer for multimodal function optimization. In: Deb, K., et al. (eds.) GECCO 2004. LNCS, vol. 3102, pp. 105–116. Springer, Heidelberg (2004)
27. van den Bergh, F., Engelbrecht, A.: A cooperative approach to particle swarm optimization. IEEE Transactions on Evolutionary Computation 8(3), 225–239 (2004)
28. Potter, M.A., de Jong, K.: A Cooperative Coevolutionary Approach to Function Optimization. In: Davidor, Y., Männer, R., Schwefel, H.-P. (eds.) PPSN 1994. LNCS, vol. 866, pp. 249–257. Springer, Heidelberg (1994)
29. Liang, J., Qin, A., Suganthan, P., Baskar, S.: Comprehensive learning particle swarm optimizer for global optimization of multimodal functions. IEEE Transactions on Evolutionary Computation 10(3), 281–295 (2006)
30. Pant, M., Thangaraj, R., Grosan, C., Abraham, A.: Hybrid differential evolution - particle swarm optimization algorithm for solving global optimization problems. In: Third International Conference on Digital Information Management (ICDIM 2008), November 2008, pp. 18–24 (2008)
31. Shelokar, P., Siarry, P., Jayaraman, V., Kulkarni, B.: Particle swarm and ant colony algorithms hybridized for improved continuous optimization. Applied Mathematics and Computation 188(1), 129–142 (2007)
32. Törn, A., Žilinskas, A.: Global Optimization. LNCS, vol. 350. Springer, Heidelberg (1989)
33. Mühlenbein, H., Schomisch, D., Born, J.: The Parallel Genetic Algorithm as Function Optimizer. Parallel Computing 17(6-7), 619–632 (1991)
34. Bäck, T., Fogel, D., Michalewicz, Z.: Handbook of Evolutionary Computation. Institute of Physics Publishing Ltd, Bristol and Oxford University Press, New York (1997)
35. Ackley, D.H.: A connectionist machine for genetic hillclimbing. Kluwer, Boston (1987)
36. Bäck, T.: Evolutionary algorithms in theory and practice. Oxford University Press, Oxford (1996)
37. Schwefel, H.P.: Numerical Optimization of Computer Models. John Wiley & Sons, Chichester (1981)
38. Dobson, I., Chiang, H.D., Thorp., J.S.: A model of voltage collapse in electric power systems. In: IEEE proceedings of 27th Conference on Decision and Control, Austin, Texas, December 1988, pp. 2104–2109 (1988)
39. Walve, K.: Modeling of power system components at severe disturbances. In: CIGRÉ paper 38-18, International conference on large high voltage electric systems (August 1986)

3

Bee Colony Optimization (BCO)

Dušan Teodorović

University of Belgrade, Faculty of Transport and Traffic Engineering, Vojvode Stepe 305
11000 Belgrade, Serbia
dusan@sf.bg.ac.rs

Abstract. Swarm Intelligence is the part of Artificial Intelligence based on study of actions of individuals in various decentralized systems. The Bee Colony Optimization (BCO) metaheuristic has been introduced fairly recently as a new direction in the field of Swarm Intelligence. Artificial bees represent agents, which collaboratively solve complex combinatorial optimization problem. The chapter presents a classification and analysis of the results achieved using Bee Colony Optimization (BCO) to model complex engineering and management processes. The primary goal of this chapter is to acquaint readers with the basic principles of Bee Colony Optimization, as well as to indicate potential BCO applications in engineering and management.

1 Introduction

Many species in the nature are characterized by swarm behavior. Fish schools, flocks of birds, and herds of land animals are formed as a result of *biological needs* to stay together. Individuals in herd, fish school, or flock of birds has a higher probability to stay alive, since predator usually assault only one individual. A collective movement characterizes flocks of birds, herds of animals, and fish schools. Herds of animals respond quickly to changes in the direction and speed of their neighbors. Swarm behavior is also one of the main characteristics of social insects (bees, wasps, ants, termites). Communication between individual insects in a colony of social insects has been well known. The communication systems between individual insects contribute to the configuration of the "collective intelligence" of the social insect colonies. The term "Swarm intelligence", that denotes this "collective intelligence" has come into use [1], [2], [3], [4].

Swarm Intelligence [4] is the part of Artificial Intelligence based on study of actions of individuals in various decentralized systems. These decentralized systems (Multi Agent Systems) are composed of physical individuals (robots, for example) or "virtual" (artificial) ones that communicate among themselves, cooperate, collaborate, exchange information and knowledge and perform some tasks in their environment.

The Bee Colony Optimization (BCO) metaheuristic [5], [6], [7], [8], [9] has been introduced fairly recently by Lučić and Teodorović as a new direction in the field of Swarm Intelligence. The BCO has been successfully applied to various engineering and management problems by Teodorović and coauthors ([10], [11], [12], [13], [14],

C.P. Lim et al. (Eds.): Innovations in Swarm Intelligence, SCI 248, pp. 39–60.
springerlink.com

[15], [16], [17]). The BCO approach is a "bottom-up" approach to modeling where special kinds of artificial agents are created by analogy with bees. Artificial bees represent agents, which collaboratively solve complex combinatorial optimization problem. The chapter presents a classification and analysis of the results achieved using BCO to model complex engineering and management processes. The primary goal of this paper is to acquaint readers with the basic principles of Bee Colony Optimization, as well as to indicate potential BCO applications in engineering and management.

2 Algorithms Inspired by Bees' Behavior in the Nature

The BCO is inspired by bees' behavior in the nature. The basic idea behind the BCO is to create the multi agent system (colony of artificial bees) capable to successfully solve difficult combinatorial optimization problems. The artificial bee colony behaves partially alike, and partially differently from bee colonies in nature. We will first describe the behavior of bees' in nature, as well as other algorithms inspired by bee s behavior. Then, we will describe a general Bee Colony Optimization algorithm and afterwards BCO applications in various engineering and management problems.

In spite of the existence of a large number of different social insect species, and variation in their behavioral patterns, it is possible to describe individual insects' as capable of performing a variety of complex tasks [18]. The best example is the collection and processing of nectar, the practice of which is highly organized. Each bee decides to reach the nectar source by following a nestmate who has already discovered a patch of flowers. Each hive has a so-called dance floor area in which the bees that have discovered nectar sources dance, in that way trying to convince their nestmates to follow them. If a bee decides to leave the hive to get nectar, she follows one of the bee dancers to one of the nectar areas. Upon arrival, the foraging bee takes a load of nectar and returns to the hive relinquishing the nectar to a food storer bee. After she relinquishes the food, the bee can (a) abandon the food source and become again uncommitted follower, (b) continue to forage at the food source without recruiting the nestmates, or (c) dance and thus recruit the nestmates before the return to the food source. The bee opts for one of the above alternatives with a certain probability. Within the dance area, the bee dancers "advertise" different food areas. The mechanisms by which the bee decides to follow a specific dancer are not well understood, but it is considered that "the recruitment among bees is always a function of the quality of the food source" [18].

Few algorithms inspired by bees' behavior appeared during the last decade (Bee System, BCO algorithm, ABC algorithm, MBO, Bees Algorithm, HBMO algorithm, BeeHive, Artificial Bee Colony, VBA algorithm). The year of publication, the names of the authors, the names of the algorithm, and the problems studied are shown in the Table 1. In a subsequent section we describe basic principles of these algorithms and we show their potential applications.

Yonezawa and Kikuchi described collective intelligence based on bees' behavior [19].

Sato and Hagiwara [20] proposed an improved genetic algorithm named *Bee System.* The proposed *Bee System* employs new operations - *concentrated crossover* and *Pseudo-Simplex Method.* By computer simulations the authors showed that the

Table 1. The algorithms inspired by bees' behavior

Year	Authors	Algorithm	Problem studied
1996	Yonezawa and Kikuchi	Ecological algorithm	Description of the collective intelligence based on bees' behavior
1997	Sato and Hagiwara	Bee System (BS)	Genetic Algorithm Improvement
2001	Lučić and Teodorović	BCO	Traveling salesman problem
2001	Abbas	MBO	Propositional satisfiability problems
2002	Lučić and Teodorović	BCO	Traveling salesman problem
2003	Lučić and Teodorović	BCO	Vehicle routing problem in the case of uncertain demand
2003	Lučić and Teodorović	BCO	Traveling salesman problem
2004	Wedde, Farooq, and Zhang	BeeHive	Routing protocols
2005	Teodorović, and Dell' Orco	BCO	Ride-matching problem
2005	Karaboga	ABC	Numerical optimization
2005	Drias, Sadeg, and Yahi	BSO	Maximum Weighted Satisfiability Problem
2005	Yang	Virtual Bee Algorithm (VBA)	Function optimizations with the application in engineering problems
2005	Benatchba, Admane, and Koudil	MBO	Max-Sat problem
2006	Teodorović, Lučić, Marković, and Dell' Orco	BCO	Traveling salesman problem and a routing problems in networks
2006	Chong, Low, Sivakumar, and Gay	Honey Bee Colony Algorithms	Job shop scheduling problem
2006	Pham, Soroka, Ghanbarzadeh, and Koc	Bees Algorithm	Optimization of neural networks for wood defect detection
2006	Basturk and Karaboga	ABC	Numeric function optimization
2006	Navrat	Bee Hive Model	Web search
2006	Wedde, Timm, and Farooq	BeeHiveAIS	Routing protocols
2007	Yang, Chen, and Tu	MBO	Improvement of the MBO algorithm
2007	Koudil, Benatchba, Tarabetand, and El Batoul Sahraoui	MBO	Partitioning and scheduling problems
2007	Quijano and Passino	Honey Bee Social Foraging Algorithm	Solving optimal resource allocation problems
2007	Marković, Teodorović, and Aćimović-Raspopović	BCO	Routing and wavelength assignment in all-optical networks
2007	Wedde, Lehnhoff, B.van Bonn, Bay, Becker, Böttcher, Brunner, Büscher, Fürst, Lazarescu, Rotaru, Senge, Steinbach, Yilmaz, and Zimmermann	BeeHive	Highway traffic congestion mitigation
2007	Karaboga and Basturk	ABC	Testing ABC algorithm on a set of multi-dimensional numerical optimization problems

Table 1. (*continued*)

Year	Authors	Algorithm	Problem studied
2007	Karaboga, Akay and Ozturk	ABC	Feed-forward neural networks training
2007	Afshar, Bozorg Haddada, Marin, Adams	Honey-bee mating optimization (HBMO) algorithm	Single reservoir operation optimization problems
2007	Baykasoglu, Özbakýr, and Tapkan	Artificial Bee Colony	Generalized Assignment Problem
2007	Teodorović and Šelmić	BCO	*p*-Median Problem
2008	Karaboga and Basturk	ABC	Comparison performances of ABC algorithm with the performances of other population-based techniques
2008	Fathian, Amiri, and Maroosi	Honeybee mating optimization algorithm	Cluster analysis
2008	Teodorović	BCO	Comparison performances of BCO algorithm with the performances of other Swarm Intelligence-based techniques
2009	Pham, Haj Darwish, Eldukhr	Bees Algorithm	Tuning the parameters of a fuzzy logic controller
2009	Davidović, Šelmić and Teodorović	BCO	Static scheduling of independent tasks on homogeneous multiprocessor systems

Bee System has better performance than the conventional genetic algorithm. The *Bee System* proposed by Sato and Hagiwara [20] can rather be categorized as Genetic Algorithm than Swarm Intelligence algorithm.

Abbass [21] developed the *MBO* model that is based on the marriage process in honeybees. The model simulates the evolution of honeybees. The author started with a solitary colony (single queen without a family) to the emergence of an eusocial colony (one or more queens with a family). The model is applied to a fifty propositional satisfiability problems (SAT) with 50 variables and 215 constraints. The proposed MBO approach was very successful on a group of fifty hard 3-SAT problems.

Wedde et al [22] developed the *BeeHive* algorithm that is also based on honeybee behavior. The authors introduced the concept of foraging regions. Each foraging region has one representative node. There are two types of agents within the BeeHive algorithm: short distance bee agents and long distance bee agents. Short distance bee agents collect and disseminate information in the neighborhood, while long distance bee agents collect and disseminate information to typically all nodes of a network.

Karaboga [23] developed the *Artificial Bee Colony* (*ABC*) algorithm. Karaboga and Basturk, and Karaboga et. [24], [25], [26] further improved and applied the *ABC* algorithm to various problems. The authors created colony of artificial bees composed of the following agents: *employed bees* (a bee flying to the food source), *onlookers* (a bee waiting on the dance area for making decision to choose a food source) and *scouts* (a bee performing random search). In the *ABC* algorithm, half of the colony consists of employed bees. The second part of the colony is composed of onlookers. Every food source could be occupied by only one employed bee. The employed bee without food source becomes a scout. The *ABC* algorithm performs search in cycles. Each cycle consists of the following three steps: (a) Employed bees fly to the food sources, collect the nectar and return to the hive. In the hive we measure their nectar amounts; (b) Information on collected nectar amounts are on a disposal to all artificial bees. Based on this information, the onlookers select the food sources; (c) Chosen bees that become scout bees fly to the possible food sources. In the *ABC* algorithm, the initial population of the solutions is generated randomly. In the subsequent cycles, the employed bees, and the onlooker bees probabilistically create a modifications on the initial solutions. Karaboga and Basturk [24] compared the performances of the *ABC* algorithm with the performances of the PSO, PS-EA and GA. Karaboga and Basturk [24] concluded, "that the proposed algorithm has the ability to get out of a local minimum and can be efficiently used for multivariable, multimodal function optimization". Karaboga et al. [25] also used the ABC algorithm to train feed-forward artificial neural networks. The authors compared performances of the ABC algorithm with the back propagation algorithm and the genetic algorithm. Performed experiments showed that the *ABC* algorithm could be good addition to the existing algorithms for feed-forward neural networks training.

Drias et al. [27] studied Maximum Weighted Satisfiability Problem. They proposed the *Bees Swarm Optimization* (*BSO*) algorithm. The authors tested their approach on the well-known benchmark problems. The *BSO* outperformed other evolutionary algorithms especially AC-SAT, an ant colony algorithm for SAT.

Yang et al. [28] developed the *Virtual Bee Algorithm* (*VBA*) to solve the function optimizations with the application in engineering problems. The simulations of the optimization of De Jong's test function and Keane's multi-peaked bumpy function showed that the *VBA* is usually as effective as genetic algorithms.

Benatchba et al. [29] applied the *MBO* algorithm to the Max-Sat problem.

Chong et al. [30] applied honey bees foraging model to the job shop scheduling problem. The authors presented experimental results comparing the proposed honeybee colony approach with existing approaches such as ant colony and tabu search. The experimental results showed that the performance of the algorithm is equivalent to ant colony algorithms,

Pham et al. [31], [32] proposed population-based search algorithm called the *Bees Algorithm* (*BA*). This algorithm also mimics the food foraging behavior of honeybees. The algorithm performs a neighborhood search combined with random search.

Navrat [33] presented a new approach to web search, based on a beehive metaphor. The author proposed a modified model of a beehive. The proposed model is simple, and it describes some of the processes that take place in web search.

Wedde et al. [34] developed a novel security framework, which is inspired by the principles of Artificial Immune Systems (AIS), for Nature inspired routing protocols.

Yang et al. [35] proposed a faster Marriage in Honey Bees Optimization (FMBO) algorithm with global convergence. By the proposed approach, the computation process becomes easier and faster. The global convergence characteristic of FMBO is also proved by using the Markov Chain theory.

Koudil et al. [36] studied partitioning and scheduling in the design of embedded systems. The authors applied Marriage in honey-Bees Optimization algorithm (MBO).

Quijano and Passino [37], [38] developed the *Honey Bee Social Foraging Algorithm*. The proposed algorithm was successfully applied to the optimal resource allocation problems.

Wedde et al. [39] proposed decentralized multi-agent approach (termed *BeeJamA)* on multiple layers for car routing. The proposed approach is based on the *BeeHive* algorithm.

Afshar et al. [40] applied Honey-bee mating optimization (*HBMO*) algorithm to the single reservoir operation optimization problems.

Baykasoglu et al. [41] made an excellent survey of the algorithms inspired by bees' behavior in the nature. The authors described the Artificial Bee Colony algorithm, and presented an artificial bee colony algorithm to solve Generalized Assignment Problem GAP.

Fathian et al. [42] applied algorithm inspired by bees' behavior in cluster analysis. The authors proposed a two-stage method. They used self-organizing feature maps (SOM) neural network to determine the number of clusters. In the second step, the authors used honeybee mating optimization algorithm based on K- means algorithm to find the final solution.

Pham et al. [43] used the *Bees Algorithm* to tune the parameters of a fuzzy logic controller. The controller was developed to stabilize and balance an under-actuated two-link acrobatic robot (ACROBOT) in the upright position. Simulation results showed that using the *Bees Algorithm* to optimize the membership functions of the fuzzy logic system enhanced the controller performance.

3 Bee Colony Optimization (BCO) Algorithm

Lučić and Teodorović [5], [6], [7], [8] were among first who used basic principles of collective bee intelligence in solving combinatorial optimization problems. The BCO is a population-based algorithm. Population of *artificial bees* searches for the optimal solution. Artificial bees represent agents, which collaboratively solve complex combinatorial optimization problems. Every artificial bee generates one solution to the problem. The algorithm consists of two alternating phases: *forward pass* and *backward pass*. In each forward pass, every artificial bee is exploring the search space. It applies a predefined number of moves, which construct and/or improve the solution, yielding to a new solution. Having obtained new partial solutions, the bees go again to the nest and start the second phase, the so-called backward pass. In the backward pass, all artificial bees share information about their solutions.

Let us consider Traveling Salesman Problem as an example. When solving the TSP problem by the BCO algorithm, we decompose the TSP problem into stages. In each stage, a bee chooses a new node to be added to the partial Traveling Salesman tour created so far (Figure 1).

In nature, bees would perform a dancing ceremony, which would notify other bees about the quantity of food they have collected, and the closeness of the patch to the nest. In the BCO search algorithm, the artificial bees publicize the quality of the solution, i.e. the objective function value. During the backward pass, every bee decides with a certain probability whether to abandon the created partial solution and become again uncommitted follower, or dance and thus recruit the nestmates before

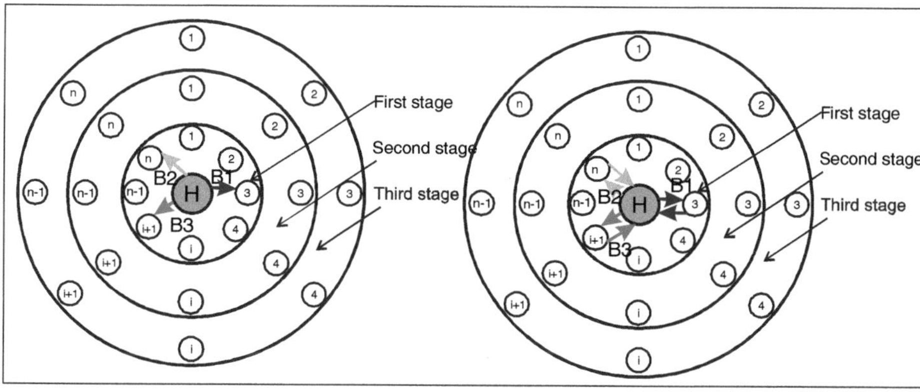

Fig. 1. First forward pass and the first backward pass

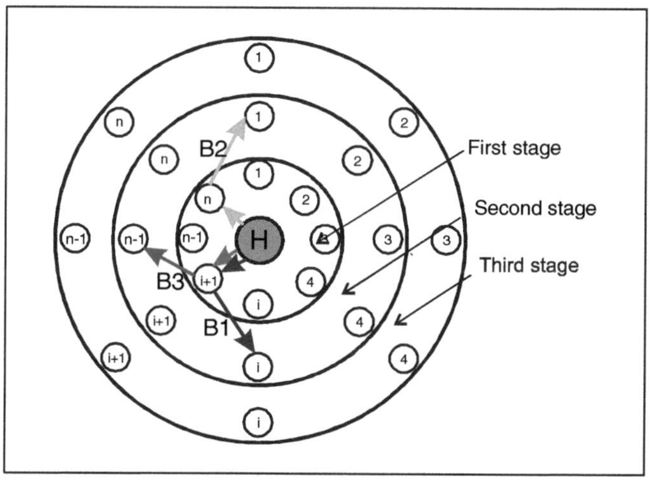

Fig. 2. Second forward pass

returning to the created partial solution (bees with higher objective function value have greater chance to continue its own exploration). Every follower, choose a new solution from recruiters (Figure 3) by the roulette wheel (better solutions have higher probability of being chosen for exploration).

During the second forward pass (Figure 2), bees expand previously created partial solutions, by a predefined number of nodes, and after that perform again the backward pass and return to the hive. In the hive, bees again participate in a decision making process, make a decision, perform third forward pass, etc. The two phases of the search algorithm, forward and backward pass, are performed *iteratively*, until a stopping condition is met. The possible stopping conditions could be, for example, the maximum total number of forward/backward passes, the maximum total number of forward/backward passes without the improvement of the objective function, etc.

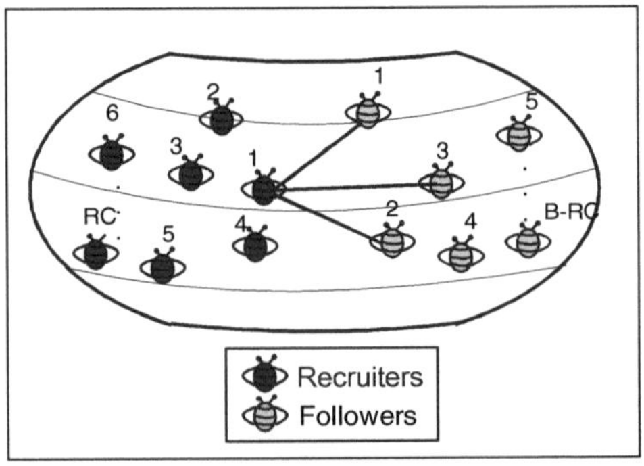

Fig. 3. Recruiting of uncommitted followers

The algorithm parameters whose values need to be set prior the algorithm execution are as follows:

B - The number of bees in the hive
NC - The number of constructive moves during one forward pass

In the beginning of the search, all the bees are in the hive. The following is the pseudocode of the BCO algorithm:

1. Initialization: every bee is set to an empty solution;
2. For every bee do the forward pass:
   a) Set $k = 1$; //counter for constructive moves in the forward pass;
   b) Evaluate all possible constructive moves;
   c) According to evaluation, choose one move using the roulette wheel;
   d) $k = k + 1$; If $k \leq NC$ Go To step b.
3. All bees are back to the hive; // backward pass starts;
4. Sort the bees by their objective function value;
5. Every bee decides randomly whether to continue its own exploration and become a recruiter, or to become a follower (bees with higher objective function value have greater chance to continue its own exploration);
6. For every follower, choose a new solution from recruiters by the roulette wheel;
7. If the stopping condition is not met Go To step 2;
8. Output the best result.

3.1 Constructive and Improving BCO Variants

A combinatorial optimization algorithm could be of constructive or improving type. Constructive approaches start from scratch. Within these approaches the analyst construct a solution step by step. When doing this, we usually apply some problem specific heuristics. On the other hand, improving approaches begin from a complete solution. The complete solution (possible a feasible one) is typically generated randomly or by some heuristics. By perturbing that solution, we try to improve it. The examples of such techniques are Simulated Annealing, or Tabu Search. Until now, the BCO algorithms in the literature have been constructive. Todorović et al. [16] developed a bee colony approach for the nurse rostering problem. Their approach is the first one that allows both constructive and improving steps to be applied and combined together.

3.2 The Artificial Bees and Approximate Reasoning

Artificial Bees confront few decision-making problems while searching for the optimal solution. The next are bees' choice dilemmas: (a) What is the next solution component to be added to the partial solution? (b) Should the partial solution be discarded or not? The greater part of the choice models in the literature, are based on random utility modeling concepts. These approaches are highly rational. They are based on assumptions that decision makers have perfect information processing capabilities and always act in a rational way (trying to maximize utility). In order to present an alternative modeling approach, researchers started to use less normative theories. The basic concepts of Fuzzy Set Theory, linguistic variables, approximate reasoning, and computing with words have more sympathy for uncertainty, imprecision, and linguistically expressed observations. Following these ideas, Teodorović and Dell'Orco [10], [14] started from the assumption that the quantities perceived by artificial bees are "fuzzy". In other words, artificial bees could also use approximate reasoning and rules of fuzzy logic in their communication and acting. When adding the solution component to the current partial solution during the forward pass, a specific bee could perceive a specific solution component as 'less attractive', 'attractive', or 'very attractive'. We also assume that an artificial bee can perceive a specific attribute as 'short', 'medium' or 'long' (Figure 4), 'cheap', 'medium', or 'expensive', etc. The approximate reasoning algorithm for calculating the solution component attractiveness consists of the rules of the following type:

If the attributes of the solution component are VERY GOOD
Then the considered solution component is VERY ATTRACTIVE

The main advantage of using the approximate reasoning algorithm for calculating the solution component attractiveness is that it is possible to calculate solution component attractiveness even if some of the input data were only approximately known.

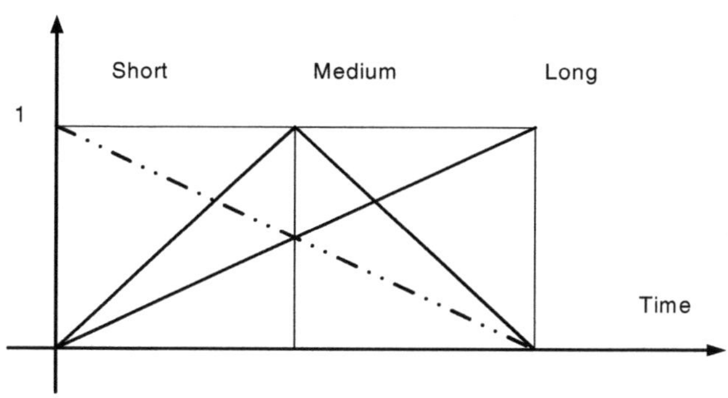

Fig. 4. Fuzzy sets describing time

4 BCO Applications

4.1 Solving the Traveling Salesman Problem by BCO

The main goal of Lučić and Teodorović [5], [6], [7], [8] research was not to develop a new heuristic algorithm for the traveling salesman problem but to explore possible applications of Swarm Intelligence (particularly collective bee intelligence) in solving complex engineering and control problems. The traveling salesman problem is only an illustrative example, which shows the characteristics of the proposed concept. Lučić and Teodorović [5], [6], [7], [8] tested the Bee Colony Optimization approach on a large number of numerical examples. The benchmark problems were taken from the following Internet address: http://www.iwr.uni-heidelberg.de/iwr/comopt/software/TSPLIB95/tsp/. The following problems were considered: Eil51.tsp, Berlin52.tsp, St70.tsp, Pr76.tsp, Kroa100.tsp and a280.tsp. All tests were run on an IBM compatible PC with PIII processor (533MHz). The results obtained are given in Table 2.

Table 2. TSP benchmark problems: The results obtained by the BCO algorithm

Problem name	Optimal value (O)	The best value obtained by the BCO (B)	$\frac{(B-O)}{O}$ (%)	CPU (sec)
Eil51	428.87	428.87	0	29
Berlin52	7544.366	7544.366	0	0
St70	677.11	677.11	0	7
Pr76	108159	108159	0	2
Kroa100	21285.4	21285.4	0	10
Eil101	640.21	640.21	0	61
Tsp225	3859	3899.9	1.06%	11651
A280	2586.77	2608.33	0.83%	6270
Pcb442	50783.55	51366.04	1.15%	4384
Pr1002	259066.6	267340.7	3.19%	28101

The solution of the benchmark problem Tsp.225 obtained by the BCO algorithm is shown in Figure 5.

We can see from the Table 2 that the proposed BCO produced results of a very high quality. The BCO was able to obtain the objective function values that are very close to the optimal values of the objective function. The times required to find the best solutions by the BCO are very low. In other words, the BCO was able to produce "very good" solutions in a "reasonable amount" of computer time.

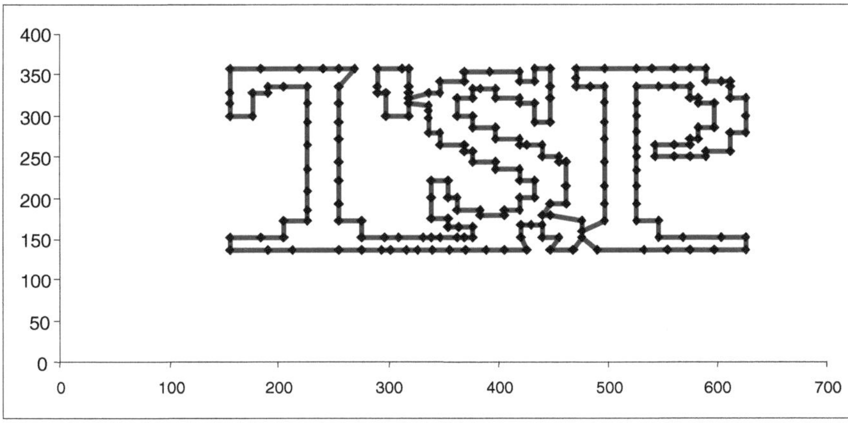

Fig. 5. Solution of the benchmark problem Tsp.225 obtained by the BCO algorithm

4.2 Solving the Ride-Matching Problem by the BCO

Urban road networks in many countries are severely congested, resulting in increased travel times, increased number of stops, unexpected delays, greater travel cost, inconvenience to drivers and passengers, increased air pollution, noise level and number of traffic accidents. Increasing traffic network capacities by building more roads is enormously costly as well as environmentally destructive. More efficient usage of the existing supply is vital in order to sustain the growing travel demand. Ridesharing is one of the widely used Travel Demand Management (TDM) techniques. Within this concept, two or more persons share vehicle when traveling from few origins to few destinations. The operator of the system must posses the following information regarding trips planned for the next week: (a) Vehicle capacity (2, 3, or 4 persons); (b) Days in the week when person is ready to participate in ride-sharing; (c) Trip origin for every day in a week; (d) Trip destination for every day in a week; (e) Desired departure and/or arrival time for every day in a week. The ride-matching problem considered by Teodorović and Dell'Orco [10], [14] could be defined in the following way: Make routing and scheduling of the vehicles and

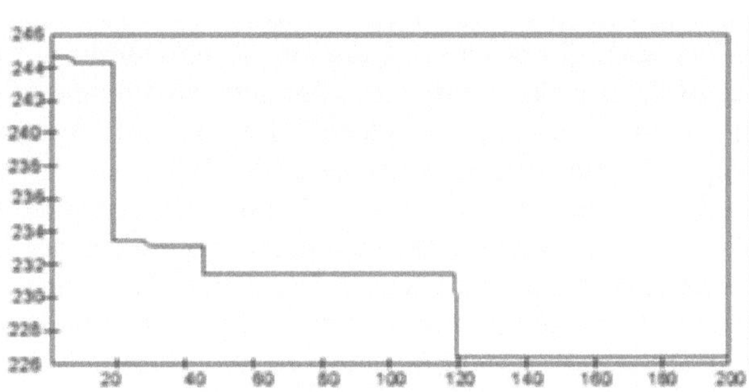

Fig. 6. Changes of the best-discovered objective function values.

passengers for the whole week in such a way to minimize the total distance traveled by all participants. Teodorović and Dell'Orco [10], [14] developed BCO based model for the ride-matching problem. The authors tested the proposed model in the case of ridesharing demand from Trani, a small city in the southeastern Italy. They collected the data regarding 97 travelers demanding for ridesharing, and assumed, for sake of simplicity, that the capacity is 4 passengers for all their cars. Changes of the best discovered objective function values are shown in Figure 6.

4.3 Routing and Wavelength Assignment in All-Optical Networks Based on the BCO

The BCO metaheuristic has been successfully tested [12] in the case of the Routing and Wavelength Assignment (RWA) in All-Optical Networks. This problem is, by its nature similar to the traffic assignment problem. The results achieved, as well as experience gained when solving the RWA problem could be used in the future research of the traffic assignment problem.

Let us briefly describe the RWA problem. Every pair of nodes in optical networks is characterized by a number of requested connections. The total number of established connections in the network depends on the routing and wavelength assignment procedure.

Routing and wavelength assignment (RWA) problem in all-optical networks could be defined in the following way: Assign a path through the network and a wavelength on that path for each considered connection between a pair of nodes in such a way to maximize the total number of established connections in the network.

Marković et al. [12] proposed the BCO heuristic algorithm tailored for the RWA problem. They called the proposed algorithm the BCO-RWA algorithm. The authors created the artificial network shown in the Figure 7. The node depicted by the square in the Figure 7 represents hive. At the beginning of the search process all artificial agents are located in the hive. Bees depart from the hive and fly through the artificial network from the left to the right. Bee's trip is divided into stages. Bee chooses to visit one artificial node at every stage. Each stage represents the collection of all considered origin-destination pairs. Each artificial node is comprised of an origin and destination linked by a number of routes. Lightpath is a route chosen by bee agent. Bee agent's entire flight is collection of established lightpaths. The authors determined in advance the number of bees B and the number of iterations I.

During forward pass every bee visits n stages (bee tries to establish n new lightpaths). In every stage a bee chooses one of the previously not visited artificial nodes. Sequence of the n visited artificial nodes generated by the bee represents one partial solution of the problem considered. Bee is not always successful in establishing lightpath when visiting artificial node. Bee's success depends on the wavelengths' availability on the specific links. In this way, generated partial solutions differ among themselves according to the total number of established lightpaths.

After forward pass, bees perform backward pass, i.e. they return to the hive. The number of nodes n to be visited during one forward pass is prescribed by the analyst at the beginning of the search process, such that $n<<m$, where m is the total number of requested lightpaths.

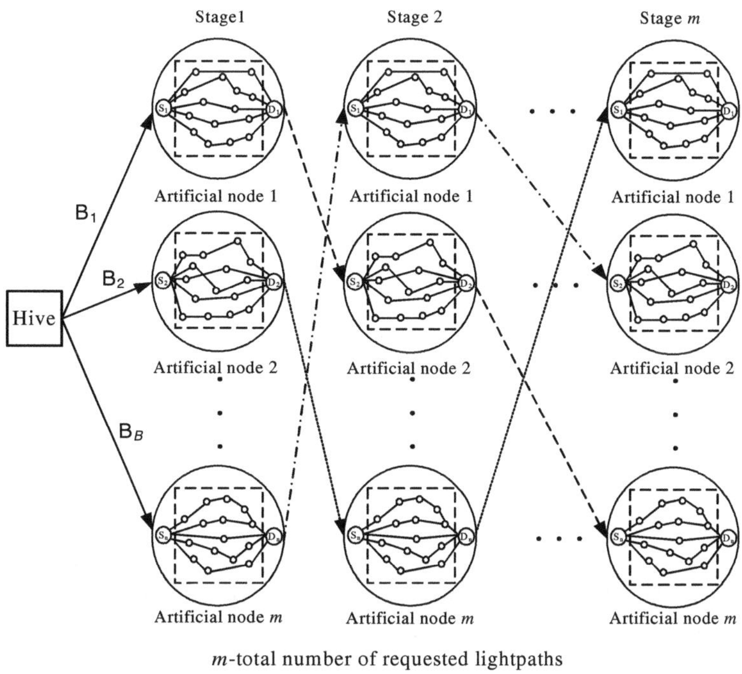

Fig. 7. Artificial network

Probability *p* that specific unvisited artificial node will be chosen by the bee equals $1/n_{tot}$, where n_{tot} is the total number of unvisited artificial nodes. By visiting specific artificial node in the network shown in Figure 7 bees attempt to establish the requested lightpath between one real source-destination node pair in optical network. Let us assume that the specific bee decided to consider the lightpath request between the source node *s* and the destination node *d*. In the next step, it is necessary to choose the route and to assign an available wavelength along the route between these two real nodes. In this paper, we defined for every node pair (*s, d*), the subset R^{sd} of allowed routes that could be used when establishing the lightpath. We defined these subsets by using the *k* shortest path algorithm. We calculated for every of the *k* alternative routes the bee's utility when choosing the considered route. The shorter the chosen route and the higher the number of available wavelengths along the route, the higher the bee's utilities are. We define the bee's utilities V_r^{sd} when choosing the route *r* between the node pair (*s, d*) in the following way:

$$V_r^{s,d} = \left\{ a \frac{1}{h_r - h_{r\min} + 1} + (1-a) \frac{W_r}{W_{\max}} \right. \tag{1}$$

where:

r – the route ordinary number for a node pair, $r = 1, 2, \ldots, k,\ r \in \{R^{sd}\}$

h_r – the route length expressed in the number of physical hops,

$h_{r\min}$ – the length of the shortest route r,
W_r – the number of available wavelengths along the route r,
$W_{\max} = \max_{r \in R^{sd}} \{W_r\}$ – the maximum number of available wavelengths among all the routes $r \in R^{sd}$
a – weight (importance of the criteria), $0 \le a \le 1$

Bees decide to choose a physical route in optical network in a random manner. Inspired by the Logit model, Marković et al. [12] assumed that the probability p_r^{sd} of choosing route r in the case of origin-destination pair (s, d) equals:

$$p_r^{sd} = \begin{cases} \dfrac{e^{V_r^{sd}}}{\sum_{i=1}^{|R^{sd}|} e^{V_i^{sd}}} & \forall r \in R^{sd} \quad and \quad W_r \rangle 0 \\ 0 & \forall r \in R^{sd} \quad and \quad W_r = 0 \end{cases} \tag{2}$$

where $|R^{sd}|$ is the total number of available routes between pair of nodes (s, d). The route r is availlable if there is at least one available wavelength on all links that belong to the route r.

In the hive every bee makes the decision about abandoning the created partial solution or expanding it in the next forward pass. The authors assumed that every bee can obtain the information about partial solution quality created by every other bee. They calculated the probability that the bee b will at beginning of the $u + 1$ forward pass use the same partial tour that is defined in forward pass u in the following way:

$$p_b = e^{-\frac{C_{\max} - C_b}{u}} \tag{3}$$

where:

C_b - the total number of established lightpaths from the beginning of the search process by the b-th bee
C_{max} - the maximal number of established lightpaths from the beginning of the search

process by any bee
u - ordinary number of forward pass, $u = 1,2,\ldots,U$

Marković et al. [12] calculated the probability p_P that the P-th advertised partial solution will be chosen by any of the uncomitted follower using the following relation:

$$p_P = \frac{e^{C_P}}{\sum_{p=1}^{P} e^{C_p}} \tag{4}$$

where C_P is the total number of the established lightpaths in the case of the P-th advertised partial solution.

The BCO-RWA algorithm was tested on a few numerical examples. The authors formulated corresponding Integer Linear Program (ILP) and discovered optimal solutions for the considered examples. In the next step, they compared the BCO-RWA results with the optimal solution. The comparison for the considered network is shown in the Table 3.

Table 3. The results comparison

Total number of requested light-paths	Number of wave-lengths	Number of established lightpaths		CPU time [s]		Relative error [%]
		ILP	BCO-RWA	ILP	BCO-RWA	
28	1	14	14	4	4.33	0
	2	23	23	94	4.58	0
	3	27	27	251	4.68	0
	4	28	28	313	4.66	0
31	1	15	14	4	4.73	6.67
	2	25	25	83	5.00	0
	3	30	30	235	5.19	0
	4	31	31	1410	5.21	0
34	1	15	14	14	5.19	6.67
	2	27	26	148	5.50	3.70
	3	33	33	216	5.64	0
	4	34	34	906	5.64	0
36	1	16	15	23	5.64	6.25
	2	27	26	325	6.09	3.70
	3	34	34	788	6.11	0
	4	36	36	1484	6.13	0
38	1	17	16	16	5.67	5.88
	2	28	27	247	6.09	3.57
	3	35	35	261	6.23	0
	4	38	38	1773	6.33	0
40	1	17	16	31	6.00	5.88
	2	28	27	491	6.28	3.57
	3	35	35	429	6.61	0
	4	40	40	1346	6.67	0

We can see from the Table 3 that the proposed BCO-RWA algorithm has been able to produce optimal, or a near-optimal solutions in a reasonable amount of computer time.

4.4 Scheduling Independent Tasks by the BCO

Davidović et al. [17] studied the problem of static scheduling of independent tasks on homogeneous multiprocessor systems. The studied problem is solved by the BCO.

The authors considered the following problem. Let $T = \{1,2,...,n\}$ be a given set of independent tasks, and $P = \{1,2,...m\}$ set of identical processors. The processing time of task i (i = 1,2,…,n) is denoted by l_i. All tasks are mutually independent and each task can be scheduled to any processor. All given tasks should be executed. Task should be scheduled to exactly one processor and processors can execute one task at a time. The goal is to find scheduling of tasks to processors in such a way as to minimize the completion time of all tasks (the so called *makespan*).

The considered scheduling problem could be graphically represented by Gantt diagram (Figure 8).

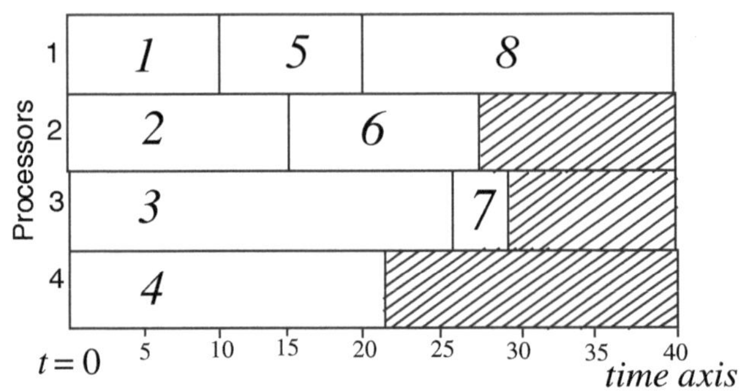

Fig. 8. Gantt diagram: Scheduled tasks to processors

The horizontal axis in the diagram represents time. The rectangulars in the Gantt diagram represent tasks. The starting time of a task is determined by the completion times of all tasks already scheduled to the same processor. The total completion time for the schedule shown in the Figure 8 equals 40 time units (the completion time of task 8 scheduled to processor 1). Any schedule that has completion time less than 40 time units is considered better. The goal is to discover the schedule of tasks to processors that has shortest completion time.

Let us briefly describe the main results achieved by Davidović et al. [17]. The authors decomposed considered problem in stages. The first task to be chosen represents the first stage, the second task to be chosen represents the second stage, the third task represents the third stage, etc. They denoted by p_i the probability that specific bee chooses task i. The probability of choosing task i equals:

$$p_i = \frac{l_i}{\sum_{k=1}^{K} l_k} \qquad i=1,2,\ldots,n \tag{5}$$

where:

l_i - processing time of the i-th task;

K - the number of "free" tasks (not previously chosen).

Obviously, tasks with a higher processing time have a higher chance to be chosen. Using relation (6) and a random number generator, we schedule task to bee. Let us denote by p_j the probability that specific bee chooses processor j. Davidović et al.[17] assumed that the probability of choosing processor j equals:

$$p_j = \frac{V_j}{\sum_{k=1}^{m} V_k} \qquad j=1,2,\ldots,m \tag{6}$$

where:

$$V_j = \frac{\max F - F_j}{\max F - \min F} \qquad j=1,2,\ldots,m \tag{7}$$

F_j - running time of processor j based on tasks already scheduled to it;
max F- maximum over all processors running times;
min F - minimum over all processors running times.

Processors with a lower value of the running times have a higher chance to be chosen. Using relation (6) and a random number generator, we schedul processor to previously chosen task. In total, B bees choose $B*NC$ task-processor pairs after the first forward pass. After scheduling tasks to processors we update processors' running times.

All bees return to the hive after generating the partial solutions. All these solutions are then evaluated by all bees. (The latest time point of finishing the last task at any processor characterizes every generated partial solution). Let us denote by C_b (b=1, 2,..., B) the latest time point of finishing the last task at any processor in the partial solution generated by the b-th bee. We denote by O_b normalized value of the time point C_b, i.e.:

$$O_b = \frac{C_{\max} - C_b}{C_{\max} - C_{\min}}, \quad b = 1,2,\ldots, B \tag{8}$$

where C_{min} and C_{max} are respectively the smallest and the largest time point among all time points produced by all bees. The probability that b-th bee (at the beginning of the new forward pass) is loyal to the previously discovered partial solution is calculated in this paper in the following way:

$$p_b^{u+1} = e^{-\frac{O_{\max} - O_b}{u}} \qquad b = 1,2,\ldots B \tag{9}$$

where u is the ordinary number of the forward pass.

Within the dance area the bee-dancers (recruiters) "advertise" different partial solutions. We have assumed in this paper that the probability the recruiter b's partial solution will be chosen by any uncommitted bee equals:

$$p_b = \frac{O_b}{\sum_{k=1}^{R} O_k} \tag{10}$$

where:

O_k - objective function value of the k-th advertised solution;

R - the number of recruiters.

Using relation (10) and a random number generator, every uncommitted follower join one bee dancer (recruiter). Recruiters fly together with a recruted nestmates in the next forward pass along the path discovered by the recruiter. At the end of this path all bees are free to independently search the solution space.

The proposed algorithm was tested on a various test problems. We denote respectively by *NT*, and *NP* the number of tasks and the number of processors. The problem parameters range from instances with NT = 10 up to the instances with NT = 50. In all cases we set NP = 4. The algorithm parameters whose values need to be set prior the algorithm execution are as follows: The total number of bees B engaged in the search process was equal to 10; The number of moves (generated task-processor pairs) *NC* during one forward pass was equal to 1; The number of iterations I within one run was equal to 100.

The authors compared the obtained BCO results with the optimal solution. The comparison results are shown in the Table 1. Within the table, BCO represents objective function value obtained by the BCO algorithm; OPT denotes the optimal makespan obtained by using ILOG AMPL and CPLEX 11.2 optimization software; CPU time shows the time required by BCO algorithm to obtain the optimal solution; I stands for the number of iteration until optimal solution was reached.

Table 4. The comparison of the BCO results with objective function optimal values for medium problems (*NT*=50, *NP*=4)

Test problem	BCO	OPT	BCO time (sec)	I
It50 70	212	212	1.0776	5
It50 80	196	196	0.5637	1
It50 80_1	234	234	1.1848	4
It50 80_2	337	337	1.7368	8
It50 80_3	216	216	0.8077	3
It50 80_4	276	276	0.7472	2
It50 80_5	128	128	0.8814	4
It50 80_6	167	167	1.8514	8

The BCO algorithm was able to obtain the optimal value of objective function in all test problems. The CPU times required to find the best solutions by the BCO are acceptable. All the tests were performed on AMD Sempron (tm) Processor with 1.60 GHz and 512 MB of RAM under Windows OS.

5 Conclusion

The Bee Colony Optimization is the youngest Swarm Intelligence technique. It is a meta-heuristic motivated by foraging behavior of honeybees. It represents general algorithmic framework that can be applied to various optimization problems in management, engineering, and control. This general algorithmic framework should be always *tailored* for a specific problem. The BCO approach is based on the concept of *cooperation*. Cooperation enables artificial bees to be more efficient and to achieve goals they could not achieve individually. The BCO has the capability, through the information exchange and recruiting process, to intensify the search in the promising regions of the solution space. When it is necessary, the BCO can also diversify the search. The BCO has not been widely used for solving real-life problems. There are no theoretical results in this moment that could support BCO concepts. Usually, development of various Swarm Intelligence approaches was based on experimental work in initial stage. Good experimental results should motivate researchers to try to produce some theoretical results in future research. Preliminary results have shown that the development of new models based on BCO principles (autonomy, distributed functioning, self-organizing) could probably contribute significantly to solving complex engineering, management, and control problems. In years to come, one could expect more BCO based models.

The most important aspect of future research is the mathematical justification of the BCO technique. The other interesting aspects of the future research could be bees' homogeneity (homogenous vs. heterogeneous artificial bees), various information sharing mechanisms, and various collaboration mechanisms.

Acknowledgment

This research is partially supported by the Ministry of Science of Serbia.

References

1. Beni, G.: The concept of cellular robotic system. In: Proceedings of the 1988 IEEE International Symposium on Intelligent Control, pp. 57–62. IEEE Computer Society Press, Los Alamitos (1988)
2. Beni, G., Wang, J.: Swarm intelligence. In: Proceedings of the Seventh Annual Meeting of the Robotics Society of Japan, pp. 425–428. RSJ Press, Tokyo (1989)
3. Beni, G., Hackwood, S.: Stationary waves in cyclic swarms. In: Proceedings of the 1992 International Symposium on Intelligent Control, pp. 234–242. IEEE Computer Society Press, Los Alamitos (1992)
4. Bonabeau, E., Dorigo, M., Theraulaz, G.: Swarm Intelligence. Oxford University Press, Oxford (1997)
5. Lučić, P., Teodorović, D.: Bee system: modeling combinatorial optimization transportation engineering problems by swarm intelligence. In: Preprints of the TRISTAN IV Triennial Symposium on Transportation Analysis, Sao Miguel, Azores Islands, Portugal, pp. 441–445 (2001)

6. Lučić, P., Teodorović, D.: Transportation modeling: an artificial life approach. In: Proceedings of the 14th IEEE International Conference on Tools with Artificial Intelligence, Washington, DC, pp. 216–223 (2002)
7. Lučić, P., Teodorović, D.: Computing with bees: attacking complex transportation engineering problems. Int. J. Artif. Intell. T. 12, 375–394 (2003a)
8. Lučić, P., Teodorović, D.: Vehicle routing problem with uncertain demand at nodes: the bee system and fuzzy logic approach. In: Verdegay, J.L. (ed.) Fuzzy Sets in Optimization, pp. 67–82. Springer, Heidelberg (2003b)
9. Teodorović, D.: Transport Modeling by Multi-Agent Systems: A Swarm Intelligence Approach. Transport. Plan. Techn. 26, 289–312 (2003b)
10. Teodorović, D., Dell'Orco, M.: Bee colony optimization – a cooperative learning approach to complex transportation problems. In: Advanced OR and AI Methods in Transportation. Proceedings of the 10th Meeting of the EURO Working Group on Transportation, Poznan, Poland, pp. 51–60 (2005)
11. Teodorović, D., Lučić, P., Marković, G., Dell' Orco, M.: Bee colony optimization: principles and applications. In: Reljin, B., Stanković, S. (eds.) Proceedings of the Eight Seminar on Neural Network Applications in Electrical Engineering – NEUREL 2006, University of Belgrade, Belgrade, pp. 151–156 (2006)
12. Marković, G., Teodorović, D., Aćimovic´ Raspopović, V.: Routing and wavelength assignment in all-optical networks based on the bee colony optimization. AI Commun. 20, 273–285 (2007)
13. Teodorović, D., Šelmić, M.: The BCO Algorithm For The p Median Problem. In: Proceedings of the XXXIV Serbian Operations Research Conferece. Zlatibor, Serbia (2007) (in Serbian)
14. Teodorović, D., Dell'Orco, M.: Mitigating traffic congestion: solving the ride-matching problem by bee colony optimization. Transport. Plan. Techn. 31, 135–152 (2008)
15. Teodorović, D.: Swarm Intelligence Systems for Transportation Engineering: Principles and Applications. Transp. Res. Pt. C-Emerg. Technol. 16, 651–782 (2008)
16. Todorović, N. Petrović, S., Teodorović, D.: Bee Colony Optimization for Nurse Rostering (submitted)
17. Davidović, T., Šelmić, M., Teodorović, D.: Scheduling Independent Tasks: Bee Colony Optimization Approach (submitted)
18. Camazine, S., Sneyd, J.: A Model of Collective Nectar Source by Honey Bees: Self-organization Through Simple Rules. J. Theor. Biol. 149, 547–571 (1991)
19. Yonezawa, Y., Kikuchi, T.: Ecological algorithm for optimal ordering used by collective Honey bee behavior. In: Proceedings of the Seventh International Symposium on Micro Machine and Humane Science, Nagoya, Japan, pp. 249–255 (1996)
20. Sato, T., Hagiwara, M.: Bee System: Finding Solution by a Concentrated Search. In: Proceedings of the IEEE International Conference on Systems, Man, and Cybernetics Computational Cybernetics and Simulation, Orlando, FL, USA, pp. 3954–3959 (1997)
21. Abbass, H.A.: MBO: marriage in honey bees optimization-a Haplometrosis polygynous swarming approach. In: Proceedings of the Congress on Evolutionary Computation, Seoul, South Korea, pp. 207–214 (2001)
22. Wedde, H.F., Farooq, M., Zhang, Y.: BeeHive: An efficient fault-tolerant routing algorithm inspired by honey bee behavior. In: Dorigo, M., Birattari, M., Blum, C., Gambardella, L.M., Mondada, F., Stützle, T. (eds.) ANTS 2004. LNCS, vol. 3172, pp. 83–94. Springer, Heidelberg (2004)

23. Karaboga, D.: An idea based on honey bee swarm for numerical optimization (Technical Report-Tr06, October, 2005), Erciyes University, Engineering Faculty Computer Engineering Department Kayseri/Türkiye (2005)
24. Karaboga, D., Basturk, B.: A powerful and efficient algorithm for numerical function optimization: artificial bee colony (ABC) algorithm. J. Global. Optim. 39, 459–471 (2007)
25. Karaboga, D., Akay, B., Ozturk, C.: Artificial Bee Colony (ABC) Optimization Algorithm for Training Feed-Forward Neural Networks. In: Torra, V., Narukawa, Y., Yoshida, Y. (eds.) MDAI 2007. LNCS (LNAI), vol. 4617, pp. 318–329. Springer, Heidelberg (2007)
26. Karaboga, D., Basturk, B.: On the performance of artificial bee colony (ABC) algorithm. Appl. Soft. Comput. 8, 687–697 (2008)
27. Drias, H., Sadeg, S., Yahi, S.: Cooperative Bees Swarm for Solving the Maximum Weighted Satisfiability Problem. In: Cabestany, J., Prieto, A.G., Sandoval, F. (eds.) IWANN 2005. LNCS, vol. 3512, pp. 318–325. Springer, Heidelberg (2005)
28. Yang, X.-S.: Engineering Optimizations via Nature-Inspired Virtual Bee Algorithms. In: Mira, J., Álvarez, J.R. (eds.) IWINAC 2005. LNCS, vol. 3562, pp. 317–323. Springer, Heidelberg (2005)
29. Benatchba, K., Admane, L., Koudil, M.: Using Bees to Solve a Data-Mining Problem Expressed as a Max-Sat One. In: Mira, J., Álvarez, J.R. (eds.) IWINAC 2005. LNCS, vol. 3562, pp. 212–220. Springer, Heidelberg (2005)
30. Chong, C.S., Low, M.Y.H., Sivakumar, A.I., Gay, K.L.: A Bee Colony Optimization Algorithm to Job Shop Scheduling Simulation. In: Perrone, L.F., Wieland, F.P., Liu, J., Lawson, B.G., Nicol, D.M., Fujimoto, R.M. (eds.) Proceedings of the Winter Conference, Washington, DC, pp. 1954–1961 (2006)
31. Pham, D.T., Ghanbarzadeh, A., Koc, E., Otri, S., Zaidi, M.: The Bees Algorithm - A Novel Tool for Complex Optimisation Problems. In: Proceedings of the 2nd Virtual International Conference on Intelligent Production Machines and Systems (IPROMS 2006), pp. 454–459. Elsevier, Cardiff (2006)
32. Pham, D.T., Soroka, A.J., Ghanbarzadeh, A., Koc, E.: Optimising Neural Networks for Identification of Wood Defects Using the Bees Algorithm. In: Proceedings of the IEEE International Conference on Industrial Informatics, Singapore, pp. 1346–1351 (2006)
33. Navrat, P.: Bee Hive Metaphor for Web Search. In: Rachev, B., Smrikarov, A. (eds.) Proceedings of the International Conference on Computer Systems and Technologies - CompSysTech 2006, Veliko Turnovo, Bulgaria, vol. 7, pp. IIIA.12- 1-7 (2006)
34. Wedde, H.F., Timm, C., Farooq, M.: BeeHiveAIS: A Simple, Efficient, Scalable and Secure Routing Framework Inspired by Artificial Immune Systems. In: Runarsson, T.P., Beyer, H.-G., Burke, E.K., Merelo-Guervós, J.J., Whitley, L.D., Yao, X. (eds.) PPSN 2006. LNCS, vol. 4193, pp. 623–632. Springer, Heidelberg (2006)
35. Yang, C., Chen, J., Tu, X.: Algorithm of Fast Marriage in Honey Bees Optimization and Convergence Analysis. In: Proceedings of the IEEE International Conference on Automation and Logistics, Jinan, China, pp. 1794–1799 (2007)
36. Koudil, M., Benatchba, K., Tarabetand, A.: El Batoul Sahraoui: Using artificial bees to solve partitioning and scheduling problems in codesign. Appl. Math. Comput. 186, 1710–1722 (2007)
37. Quijano, N., Passino, K.M.: Honey Bee Social Foraging Algorithms for Resource Allocation, Part I: Algorithm and Theory. In: Proceedings of the 2007 American Control Conference, New York, pp. 3383–3388 (2007a)
38. Quijano, N., Passino, K.M.: Honey Bee Social Foraging Algorithms for Resource Allocation, Part II: Application. In: Proceedings of the 2007 American Control Conference, New York, pp. 3389–3394 (2007b)

39. Wedde, H.F., Lehnhoff, S., van Bonn, B., Bay, Z., Becker, S., Böttcher, S., Brunner, C., Büscher, A., Fürst, T., Lazarescu, M., Rotaru, E., Senge, S., Steinbach, B., Yilmaz, F., Zimmermann, T.: A Novel Class of Multi-Agent Algorithms for Highly Dynamic Transport Planning Inspired by Honey Bee Behavior. In: Proceedings of the 12th IEEE International Conference on Factory Automation, Patras, Greece, pp. 1157–1164 (2007)
40. Afshar, A., Bozorg Haddada, O., Marin, M.A., Adams, B.J.: Honey-bee mating optimization (HBMO) algorithm for optimal reservoir operation. J. Frank. Instit. 344, 452–462 (2007)
41. Baykasoglu, A., Özbakýr, L., Tapkan, P.: Artificial Bee Colony Algorithm and Its Application to Generalized Assignment Problem. In: Chan, F.T.S., Tiwari, M.K. (eds.) Swarm Intelligence: Focus on Ant and Particle Swarm Optimization, pp. 113–143. Itech Education and Publishing, Vienna (2007)
42. Fathian, M., Amiri, B., Maroosi, B.: A honeybee-mating approach for cluster analysis. Int. J. Adv. Manuf. Technol. 38, 809–821 (2008)
43. Pham, D.T., Haj Darwish, A., Eldukhr, E.E.: Optimisation of a fuzzy logic controller using the Bees Algorithm. Int. J., Comp. Aid. Eng. Tech. 1, 250–264 (2009)

4

Glowworm Swarm Optimization for Searching Higher Dimensional Spaces

K.N. Krishnanand[1] and D. Ghose[2]

[1] Department of Computer Science, University of Vermont, Burligton, VT, USA
krishna@cems.uvm.edu, krishna.iisc@gmail.com
[2] Guidance, Control and Decision Systems Laboratoty (GCDSL),
Department of Aerospace Engineering, Indian Institute of Science,
Bangalore 560 012, India
dghose@aero.iisc.ernet.in

Abstract. This chapter will deal with the problem of searching higher dimensional spaces using glowworm swarm optimization (GSO), a novel swarm intelligence algorithm, which was recently proposed for simultaneous capture of multiple optima of multimodal functions. Tests are performed on a set of three benchmark functions and the *average peak-capture fraction* is used as an index to analyze GSO's performance as a function of dimension number. Results reported from tests conducted up to a maximum of eight dimensions show the efficacy of GSO in capturing multiple peaks in high dimensions. With an ability to search for local peaks of a function (which is the measure of fitness) in high dimensions, GSO can be applied to identification of multiple data clusters, satisfying some measure of fitness defined on the data, in high dimensional databases.

1 Introduction

Extracting relevant data that satisfies or optimizes some specifications, from semi-structured knowledge databases, can involve evaluation of candidate data using several parameters. This leads to the problem of searching in higher dimensional spaces. This has been addressed by researchers in the current literature. However, in practical databases, there is also a need to identify several data points that may satisfy these specifications. This is one of the major objectives of the method proposed in this chapter and is different from the usual requirement of searching for a point that is globally optimal. In particular, we address the problem of searching higher dimensional spaces using a novel swarm intelligence algorithm, called glowworm swarm optimization (GSO), which was proposed recently (Krishnanand and Ghose 2009a;b; 2008; 2006).

Goal of searching a database is to search for clusters of data that have high value of some measure of fitness defined on the data. There could be several clusters that satisfy these requirements. Such data clusters will have traces of influence on neighboring data, which can be used as indicators of the existence of high value data clusters. For example, database of traces of minerals on the earth's crust reveal that mineral concentrations

C.P. Lim et al. (Eds.): Innovations in Swarm Intelligence, SCI 248, pp. 61–75.
springerlink.com

are good indicators of large concentration of minerals in a close neighborhood. Also, these concentrations need not be at a single location but, as is usual, may be located at several different geographical locations. The existence of high concentrations need not depend only on the concentration level of the trace mineral (which could be very low) but also on other parameters such as traces of other elements and their reactive residues. Thus, the search algorithm essentially should use evaluation of data through several parameters to search for these high valued data clusters. This problem is similar to searching for local peaks of a function (which is the measure of fitness) in high dimensions. Note that the function profile can be rather chaotic and hence standard gradient based algorithms are unlikely to work. In the mineral detection example, such a search algorithm can be used to determine existence of large concentrations of minerals in some geographical area.

GSO is a novel algorithm that was recently developed for simultaneous computation of multiple optima of multimodal functions. The agents in GSO are thought of as glowworms that carry a luminescence quantity called luciferin along with them. The glowworms encode the fitness of their current locations, evaluated using the objective function, into a luciferin value that they broadcast to their neighbors. The glowworm identifies its neighbors and computes its movements by exploiting an adaptive neighborhood, which is bounded above by its sensor range. Each glowworm selects, using a probabilistic mechanism, a neighbor that has a luciferin value higher than its own and moves toward it. These movements – based only on local information and selective neighbor interactions – enable the swarm of glowworms to partition into disjoint subgroups that converge on multiple optima of a given multimodal function. In the earlier works on GSO, the capability of GSO in capturing multiple optima was tested against a large class of benchmark multimodal functions (Krishnanand and Ghose 2009a), theoretical foundations involving local convergence results for a simplified GSO model were reported (Krishnanand and Ghose 2008), and real robot experiments were carried out in which two robots used GSO to localize a light source (Krishnanand 2007). However, in all the above, GSO has been confined to low dimension search spaces. In this work we apply GSO in searching higher dimensional spaces motivated by the knowledge database search application discussed above and show its applicability in solving such problems. Literature on searching for multiple solutions in high dimensional spaces is rather meager and we hope that this work will introduce new ideas in this area.

The chapter is organized as follows. Related work on search in high dimensional spaces is presented in Section 2. The GSO algorithm is briefly described in Section 3. A comparison between GSO and PSO (particle swarm optimization, which is a well-known swarm based optimization algorithm) is provided in Section 4. A summary of results from numerical experiments on a set of three benchmark multimodal functions is presented in Section 5. Concluding remarks are given in Section 6.

2 Related Work

Multimodal function optimization has been addressed extensively in the recent literature. The traditional class of problems related to this topic focused on developing algorithms to find either the global optimum or all the global optima of the given multimodal function, while avoiding local optima. However, there is another class of optimization problems which is different from the problem of finding only the global optimum of a multimodal function. The objective of this class of multimodal optimization problems is to find multiple optima having either equal or unequal function values (Brits et al. 2002; Kennedy 2000; Parsopoulos and Vrahatis 2004; Li 2004). Multi-modality in a search and optimization problem gives rise to several attractors and thereby presents a challenge to any optimization algorithm in terms of finding global optimum solutions (Singh and Deb 2006). The problem is compounded when multiple (global and local) optima are sought.

Li (Li 2004) proposed a species-based particle swarm optimization technique (SPSO) for solving multimodal function optimization problems. At each iteration step of SPSO, all the particles are sorted in decreasing order of fitness and, using a notion of species that is inspired from Goldberg and Richardson (1987), the sorted population is classified into groups according to their similarity measured by Euclidean distance, which is bounded above by a parameter r_s. For instance, all particles that are within a distance r_s from the particle with maximum fitness form the first group. The next group is formed around the particle that is next in the fitness-hierarchy and that does not belong to the first group. Proceeding in this manner, multiple species are identified and a neighborhood-best is determined for each species. These multiple adaptively formed species are then used to optimize towards multiple optima in parallel, without interference across different species. In one and two dimensions, SPSO is shown to be effective in finding multiple optima on a set of five test functions suggested by Beasley et al. (Beasley et al. 1993). SPSO was tested on the Rastrigin's function (Mühlenbein et al. 1991) up to six dimensions. However, in these tests, success rate was measured by the percentage of runs (out of 30) locating the global optimum. A success rate of 100 %, 100 %, 33.3 %, 26.7 %, and 0 % were observed for 2, 3, 4, 5, and 6 dimensions, respectively.

In Suganthan et al. (Suganthan et al. 2005), the authors presented a set of 25 benchmark functions and different evaluation criteria for real-parameter optimization. Using particle swarm optimization (PSO) (Clerc 2007) as an example, results were reported from tests conducted in 10, 30, and 50 dimensions. However, the objective in all these experiments was to capture the global optimum.

In all the above experiments on high dimensions, the focus is on locating a single global optimum. In contrast, GSO is tested for its ability to capture multiple optima in high dimensions.

3 Glowworm Swarm Optimization (GSO)

In GSO, physical entities/agents $\{i : i = 1, \ldots, n\}$ are considered that are initially randomly deployed $\{x_i(0) : x_i \in \mathbb{R}^m, i = 1, \ldots, n\}$ in the objective function space $\mathbb{R}^m$. Each agent in the swarm decides its direction of movement by the strength of the signal picked up from its neighbors. This is somewhat similar to the luciferin induced glow of a glowworm which is used to attract mates or prey. The brighter the glow, more is the attraction. Therefore, we use the glowworm metaphor to represent the underlying principles of our optimization approach. Hereafter, we refer to the agents in GSO as glowworms. However, the glowworms in GSO are endowed with other behavioral mechanisms (not found in their natural counterparts) that enable them to selectively interact with their neighbors and decide their movements at each iteration. In natural glowworms, the brightness of a glowworm's glow as perceived by its neighbor reduces with increase in the distance between the two glowworms. However, in GSO, we assume that luciferin value of a glowworm agent as perceived by its neighbor does not reduce due to distance.

Each glowworm i encodes the objective function value $J(x_i(t))$ at its current location $x_i(t)$ into a luciferin value ℓ_i and broadcasts the same within its neighborhood. Each glowworm i regards only those incoming luciferin data as useful that are broadcast by its neighbors; the set of neighbors $N_i(t)$ of glowworm i consists of those glowworms that have a relatively higher luciferin value and that are located within a dynamic decision domain whose range r_d^i is bounded above by a circular sensor range r_s $(0 < r_d^i \leq r_s)$. Each glowworm i selects a neighbor j with a probability $p_{ij}(t)$ and moves toward it. These movements, that are based only on local information, enable the glowworms to partition into disjoint subgroups, exhibit a simultaneous taxis-behavior toward, and eventually co-locate at, the multiple optima of the given objective function.

The GSO algorithm is given in the inset box. The algorithm is presented for maximization problems. However, it can be easily modified and used to find multiple minima of multimodal functions.

The algorithm starts by placing a population of n glowworms randomly in the search space so that they are well dispersed. Initially, all the glowworms contain an equal quantity of luciferin ℓ_0. Each iteration consists of a luciferin-update phase followed by a movement phase based on a transition rule.

Luciferin-update phase: The luciferin update depends on the function value at the glowworm position. During the luciferin-update phase, each glowworm adds, to its previous luciferin level, a luciferin quantity proportional to the fitness of its current location in the objective function space. Also, a fraction of the luciferin value is subtracted to simulate the decay in luciferin with time. The luciferin update rule is given by:

$$\ell_i(t+1) = (1-\rho)\ell_i(t) + \gamma J(x_i(t+1)) \tag{1}$$

where $\ell_i(t)$ represents the luciferin level associated with glowworm i at time t, ρ is the luciferin decay constant $(0 < \rho < 1)$, γ is the luciferin enhancement constant and $J(x_i(t))$ represents the value of the objective function at agent i's location at time t.

Movement phase: During the movement phase, each glowworm decides, using a probabilistic mechanism, to move toward a neighbor that has a luciferin value higher than its own. That is, glowworms are attracted to neighbors that glow brighter. For each glowworm i, the probability of moving toward a neighbor j is given by:

$$p_{ij}(t) = \frac{\ell_j(t) - \ell_i(t)}{\sum_{k \in N_i(t)} \ell_k(t) - \ell_i(t)} \tag{2}$$

where $j \in N_i(t)$, $N_i(t) = \{j : d_{ij}(t) < r_d^i(t); \ell_i(t) < \ell_j(t)\}$ is the set of neighbors of glowworm i at time t, $d_{ij}(t)$ represents the Euclidean distance between glowworms i and j at time t, and $r_d^i(t)$ represents the variable neighborhood range associated with

GLOWWORM SWARM OPTIMIZATION (GSO) ALGORITHM

Set number of dimensions $= m$
Set number of glowworms $= n$
Let s be the step size
Let $x_i(t)$ be the location of glowworm i at time t
$deploy\_agents\_randomly$;
for $i = 1$ to n do $\ell_i(0) = \ell_0$
$r_d^i(0) = r_0$
set maximum iteration number $= iter\_max$;
set $t = 1$;
while $(t \leq iter\_max)$ do:
{
for each glowworm i do: % Luciferin-update phase
$\ell_i(t) = (1-\rho)\ell_i(t-1) + \gamma J(x_i(t))$;

for each glowworm i do: % Movement-phase
{
$N_i(t) = \{j : \|x_j(t) - x_i(t)\| < r_d^i(t); \ell_i(t) < \ell_j(t)\}$;
where $\|\boldsymbol{x}\|$ is the norm of $\boldsymbol{x}$
for each glowworm $j \in N_i(t)$ do:
$p_{ij}(t) = \frac{\ell_j(t)-\ell_i(t)}{\sum_{k \in N_i(t)} \ell_k(t)-\ell_i(t)}$;
$j = select\_glowworm(\boldsymbol{p})$;
$x_i(t+1) = x_i(t) + s\left(\frac{x_j(t)-x_i(t)}{\|x_j(t)-x_i(t)\|}\right)$
$r_d^i(t+1) = \min\{r_s, \max\{0, r_d^i(t) + \beta(n_t - |N_i(t)|)\}\}$;
}
$t \leftarrow t+1$;
}

Fig. 1. *The GSO algorithm*

Table 1. Values of algorithmic parameters that are kept fixed for all the experiments

ρ	γ	β	n_t	s	ℓ_0
0.4	0.6	0.08	5	0.03	5

glowworm i at time t. Let glowworm i select a glowworm $j \in N_i(t)$ with $p_{ij}(t)$ given by (2). Then, the discrete-time model of the glowworm movements can be stated as:

$$x_i(t+1) = x_i(t) + s\left(\frac{x_j(t) - x_i(t)}{\|x_j(t) - x_i(t)\|}\right) \tag{3}$$

where $x_i(t) \in \mathbb{R}^m$ is the location of glowworm i, at time t, in the $m-$dimensional real space $\mathbb{R}^m$, $\|\cdot\|$ represents the Euclidean norm operator, and s (> 0) is the step size.

Local decision domain update rule: We associate with each agent i a local decision domain whose radial range r_d^i is dynamic in nature ($0 < r_d^i \leq r_s$). The fact that a fixed range is not used needs some justification. When the glowworms depend only on local information to decide their movements, it is expected that the number of peaks captured would be a function of the radial sensor range. In fact, if the sensor range of each agent covers the entire search space, all the agents move to the global optimum and the local optima are ignored. Since we assume that *a priori* information about the objective function (e.g., number of peaks and inter-peak distances) is not available, it is difficult to fix the decision domain range at a value that works well for different function landscapes. For instance, a chosen range r_d would work relatively better on objective functions where the minimum inter-peak distance is more than r_d rather than on those where it is less than r_d. Therefore, GSO uses an adaptive decision range in order to detect the presence of multiple peaks in a multimodal function landscape.

Let r_0 be the initial decision domain range of each glowworm (that is, $r_d^i(0) = r_0 \;\forall\, i$). To adaptively update the decision domain range of each glowworm, the following rule is applied:

$$r_d^i(t+1) = \min\{r_s, \max\{0, r_d^i(t) + \beta(n_t - |N_i(t)|)\}\} \tag{4}$$

where β is a constant parameter and n_t is a parameter used to control the number of neighbors.

The quantities $n_t, s, \ell_0, \beta, \rho$, and γ are algorithm parameters for which appropriate values have been determined based on extensive numerical experiments and are kept fixed (Table 1). Thus, only n and r_s need to be selected. A full factorial analysis is carried out in (Krishnanand and Ghose 2009a) to show that the choice of these parameters has some influence on the performance of the algorithm, in terms of the total number of peaks captured.

4 Comparison of GSO with PSO

Particle swarm optimization (PSO) is a population-based stochastic search algorithm (Clerc 2007). In PSO, a population of solutions $\{X_i : X_i \in \mathbb{R}^m, i = 1,\ldots,N\}$ is evolved, which is modeled by a swarm of particles that start from random positions in the objective function space and move through it searching for optima; the velocity V_i of each particle i is dynamically adjusted according to the *best so far* positions visited by itself (P_i) and the whole swarm (P_g). The position and velocity updates of the ith particle are given by:

$$V_i(t+1) = V_i(t) + c_1 r_1 (P_i(t) - X_i(t)) + c_2 r_2 (P_g(t) - X_i(t)) \tag{5}$$

$$X_i(t+1) = X_i(t) + V_i(t+1) \tag{6}$$

where c_1 and c_2 are positive constants, referred to as the *cognitive* and *social* parameters, respectively, r_1 and r_2 are random numbers that are uniformly distributed within the interval $[0, 1]$, $t = 1, 2, \ldots,$ indicates the iteration number. PSO uses a memory

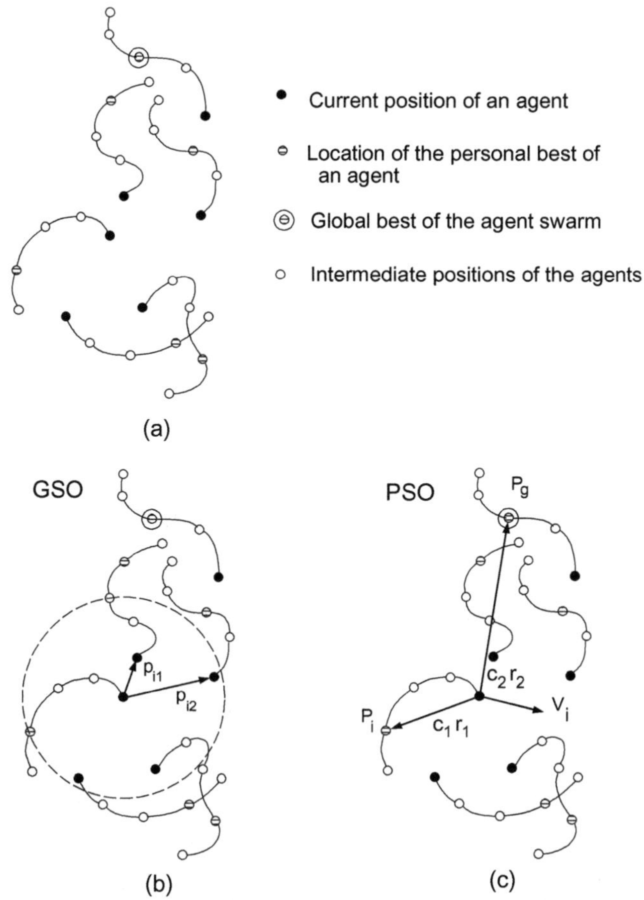

Fig. 2. (a) Paths traveled by six agents; each agent's current and personal best positions and the global best position are also shown. (b) GSO decision: the agent has four neighbors in its dynamic range of which two have higher luciferin value. Hence it has two possible directions to move with different probabilities. (c) PSO decision: the velocity of the agent is a random combination between its current velocity, the vector to the personal best location, and the vector to the global best location. (Krishnanand and Ghose 2009b)

element in the velocity update mechanism of the particles. Differently, the notion of memory in GSO is incorporated into the incremental update of the luciferin values that reflect the cumulative goodness of the path followed by the glowworms.

GSO shares a feature with PSO: In both PSO and GSO algorithms a group of particles/agents are initially deployed in the objective function space and each agent selects, and moves in, a direction based on respective position update formulae. While the directions of a particle movements in original PSO are adjusted according to its own and global best previous positions, movement directions are aligned along the line-of-sight between neighbors in GSO. In PSO, The net improvement in the objective function at the iteration t is stored in $P_g(t)$. However, in the GSO algorithm, a glowworm with the highest luciferin value in a populated neighborhood indicates its potential proximity to an optimum location. While the directions of particle movements in PSO are adjusted according to its own and global best previous positions, movement directions are aligned along the line-of-sight between neighbors in the GSO algorithm.

Figures 2(a)–2(c) (Krishnanand and Ghose 2009b) illustrate the basic concepts behind PSO and GSO and bring out the differences between them. Figure 2(a) shows the trajectories of six agents, their current positions, and the positions at which they encountered their personal best and the global best. Figure 2(b) shows the GSO decision for an agent which has four neighbors in its dynamic range of which two have higher luciferin value and thus only two probabilities are calculated. Figure 2(c) shows the PSO decision, which is a random combination between the current velocity of the agent, the vector to the personal best location, and the vector to the global best location.

In PSO, the next velocity direction and magnitude is dependent on combination of the agent's own current velocity, and randomly weighted global best vector and personal best vector. While this is implementable in a purely computational platform, implementation of this algorithm in robotics platform or a platform containing realistic agents would demand large speed fluctuations, presence of memory of the personal best position of each agent, and knowledge of the global best position which requires global communication with all other agents. In the local variant of PSO, P_g is replaced by the best previous position encountered by particles in a local neighborhood. In one of the local variants of PSO, the dynamic neighborhood is achieved by evaluating the first k nearest neighbors. However, such a neighborhood topology is also limited to computational models only and is not applicable in a realistic scenario where the neighborhood size is defined by the limited sensor range of the mobile agents.

In GSO, the next direction of movement of an agent is determined by the position of the higher luciferin carrying neighbors within a dynamic decision range and the weights are determined by the actual values of the luciferin level. Thus, in GSO the implementation is much simpler as the algorithm demands communication only with a limited number of neighbors and therefore does not require to retain personal best position information, nor does it require to collect data from all agents to determine the global best position.

Conceptually, the fact that GSO does not use the global best position or the personal best position and the fact that it uses information only from a dynamic neighbor set helps it to detect local maxima, whereas PSO gets easily attracted to the global maxima.

The above figures and explanation imply that GSO is a completely different algorithm than PSO although they have some minor commonalities.

The GSO algorithm directly addresses the issue of partitioning a swarm required by the multiple peak capture problem.

To the best of our knowledge, GSO features the most advanced method for addressing the problem of partitioning agents to all the peaks. The key to this is the adaptive local decision domain that is distinctive from the sensory range of each glowworm. By specifying a number of neighbors at the beginning of the search, each glowworm establishes its local decision domain that is smaller than its sensory range. The decisions made by glowworms in their respective local domains cause the swarm to separate into subgroups that move toward the nearest peak rather than toward a global maximum. Therefore, the adaptive local-decision domain facilitates the formation of subgroups to locate multiple peaks simultaneously.

5 Experiments

We use the following set of multimodal functions to test the efficacy of GSO in capturing multiple peaks in higher dimensions:

Rastrigin's function:

$$J_1(x) = \sum_{i=1}^{m}(10 + x_i^2 - 10\cos(2\pi x_i)) \tag{7}$$

In $J_1(x)$, $x \in \Re^m$ is a position vector in the m-dimensional search space. The Rastrigin's function (Figure 3) was first proposed by Rastrigin (Törn and Zilinskas 1989) as a 2-dimensional function and has been generalized by Mühlenbein et al. in (Mühlenbein et al. 1991). This function presents a fairly difficult problem due to its large search space and its large number of local minima and maxima.

Schwefel's function:

$$J_2(x) = \sum_{i=1}^{m}(418.9829 - x_i\sin(\sqrt{|x_i|})) \tag{8}$$

The Schwefel's function (Figure 4) presents a complexity in terms of the locations of the peaks. From Figure 4, note that the peak at (-5.24, -5.24) is interior to the search

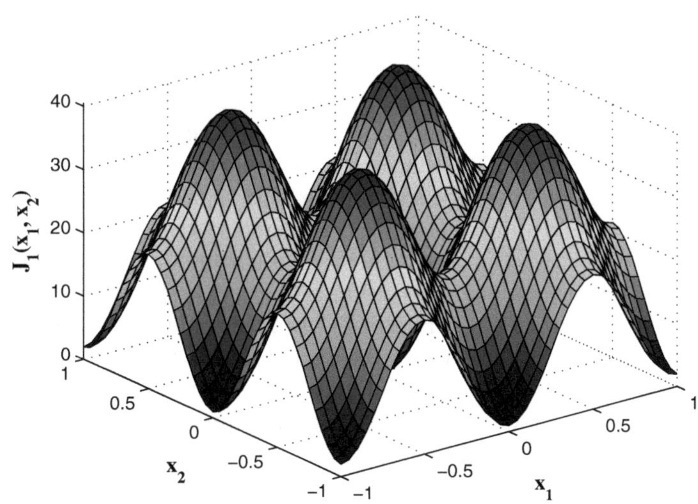

Fig. 3. Rastrigin's function in two dimensions

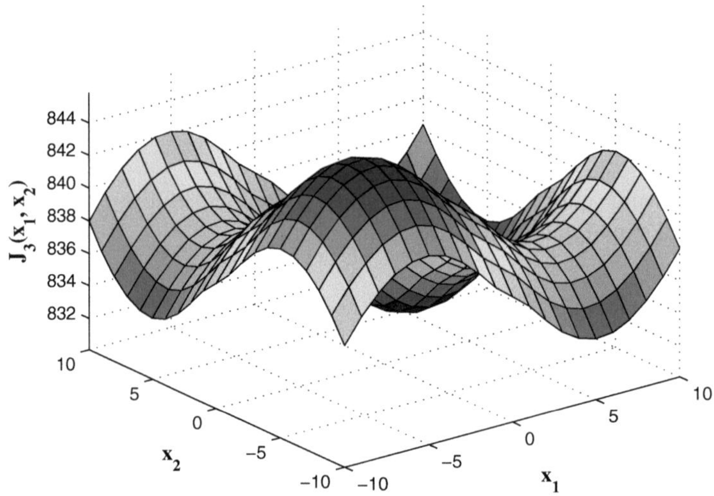

Fig. 4. Schwefel's function in two dimensions

space. However, the peaks at (-5.24,10) and (10,-5.24) are located on the edges of the search space and the peak at (10, 10) is located on a corner of the search space making it difficult for the algorithm to identify these peaks.

Equal-peaks function:

$$J_3(x) = \sum_{i=1}^{m} \sin^2(x_i) \tag{9}$$

The shape of the function profile of the Equal-peaks function (See Figure 4) is similar to that of the Rastrigin's function in the considered search space. However, their search space sizes are different from each other.

We consider a glowworm to be located at a peak when its position is within a small ϵ-distance from the peak and we assume that a peak is captured when at least three glowworms are located at the peak. The value $\epsilon = 0.03$ is chosen for tests conducted on the Rastrigin's function and the Equal-peaks function. The value $\epsilon = 0.3$ was used for tests conducted on the Schwefel's function as the considered search space was very large as compared to those of the two former functions. Peak capture results from experiments conducted in two dimensions that are shown in Figure 6 – Figure 8 can be used to visualize the basic working of GSO. Figure 6 shows the results obtained when GSO is tested on the Rastrigin's function. In Figure 6(a), a group of 200 glowworms start moving from random initial locations, partition into four subgroups, and eventually each subgroup of glowworms settles at a different peak location. Figure 6(b) shows the initial random locations and the final locations of all the glowworms. Similar test results are shown for the Schwefel's function and the Equal-peaks function in Figure 7 and Figure 8, respectively. From Figure 7(a), note that the swarm partitions into four subgroups with different population sizes. In particular, the subgroup with the highest number of

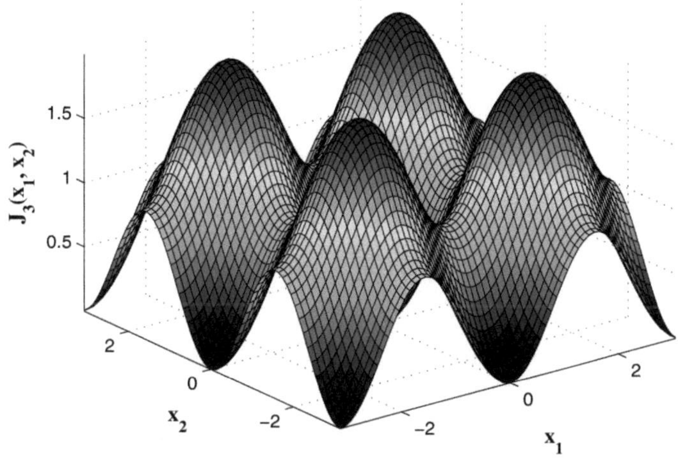

Fig. 5. Equal-peaks function in two dimensions

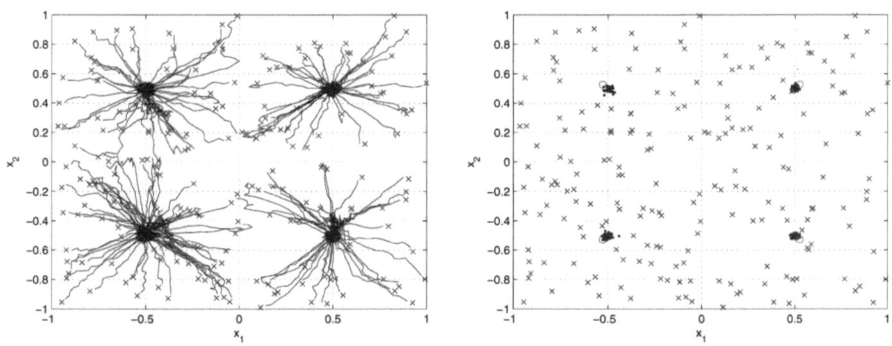

(a) (b)

Fig. 6. Rastrigin's function: (a) trajectories (200 iterations) followed by the glowworms: they start from an initial random location (cross) and they end up at one of the peaks. (b) Initial placement and final location of the glowworms.

glowworms moves to the interior peak, the subgroup with lower number of glowworms move to the two peaks on the edges of the search space, and the subgroup with the lowest number of glowworms moves to the corner peak. This explains the difficulty in locating peaks on corners and edges of the search space and is evident in the high dimensional test results that are shown later (See Figure 10). From Figure 8(a), note that all the four peaks of the Equal-peaks function are interior to the search space. Accordingly, four subgroups of approximately equal population sizes are formed. This is also true for the Rastrigin's function case (See Figure 6(a)).

Next, the capability of GSO in capturing peaks in high dimensions is tested on all the three functions J_1, J_2, and J_3. The search space size used in the experiments, the

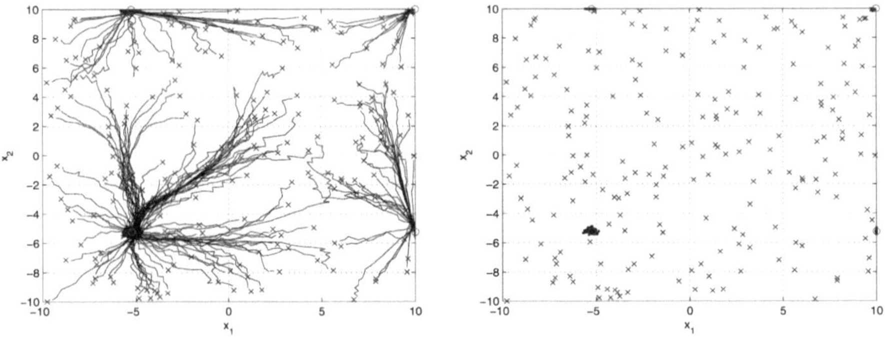

(a) (b)

Fig. 7. Schwefel's function: (a) trajectories (200 iterations) followed by the glowworms: they start from an initial random location (cross) and they end up at one of the peaks. (b) Initial placement and final location of the glowworms.

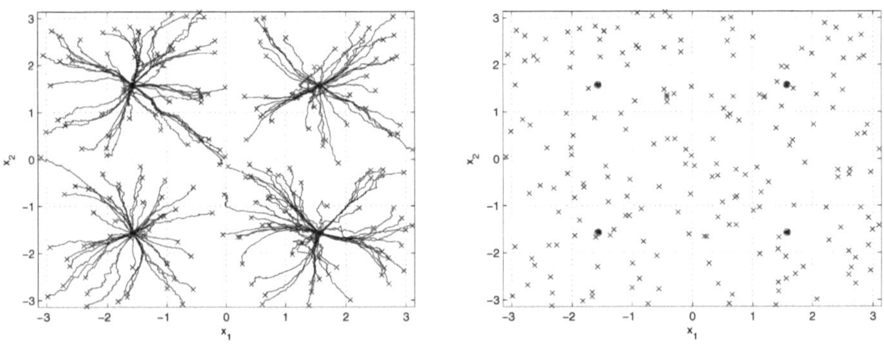

(a) (b)

Fig. 8. Equal-peaks function: (a) trajectories (200 iterations) followed by the glowworms: they start from an initial random location (cross) and they end up at one of the peaks. (b) Initial placement and final location of the glowworms.

number of peaks in m dimensions for the considered search space size, and the peak locations for $m = 2$ for each function are shown in Table 2. From the table, note that each function has 4, 8, 16, 32, 64, 128, and 256 peaks in 2, 3, 4, 5, 6, 7, and 8 dimensions, respectively. The fraction of peaks captured averaged over a set of thirty trials is used as a measure to access the performance of GSO in capturing multiple peaks at each dimension. A maximum value of n = 10,000 is used in the experiments. For each multimodal function, the tests are conducted up to a value of m at which the average peak capture fraction drops below 0.5.

Figure 9 shows a plot of average fraction of peaks captured as a function of dimension number m when Rastrigin's function is used in the experiments. Note that 100 % peak capture is obtained up to $m = 6$, followed by peak captures of 98 $\pm$ 1.3 % and 45 $\pm$ 33.8 % at $m = 7$ and 8, respectively. Figure 10 shows the test results in the case of Schwefel's function. Note that 100 % peak capture is found only in two dimensions,

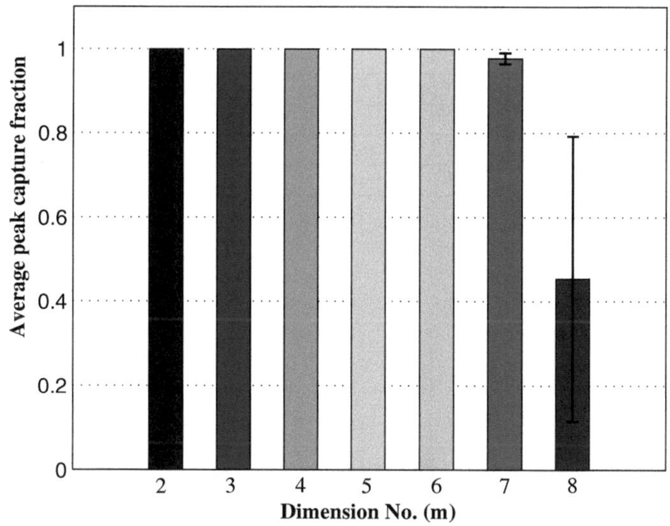

Fig. 9. Test results on Rastrigin's function: Plot of average fraction of peaks captured as a function of dimension number m (averaged over 30 trials)

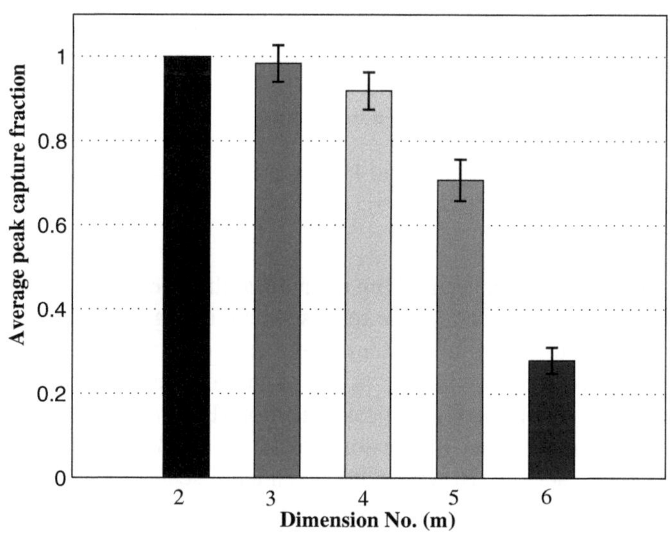

Fig. 10. Test results on Schwefel's function: Plot of average fraction of peaks captured as a function of dimension number m (averaged over 30 trials)

followed by 98.3 $\pm$ 4.3, 92 $\pm$ 4.4 %, and 70 $\pm$ 5 % at m = 3, 4, and 5, respectively. The peak capturing performance degrades from m = 6 onwards. As mentioned earlier, in the case of the Schwefel's function, for any dimension m, only one out of 2^m peaks is interior to the search space and each of the remaining peaks is located either on an

Table 2. Search space size, number of peaks, and peak locations for $m = 2$ for each multimodal function.

Function	Search space size	No. of peaks	Peak locations for $m = 2$
J_1	$\prod_{i=1}^{m}$ [-1, 1]	2^m	(±0.5268, ±0.5268)
J_2	$\prod_{i=1}^{m}$ [-10, 10]	2^m	(-5.24, -5.24)(-5.24,10), (10,-5.24), (10, 10),
J_3	$\prod_{i=1}^{m}$ [-π, π]	2^m	$(\pm\frac{\pi}{2}, \pm\frac{\pi}{2})$

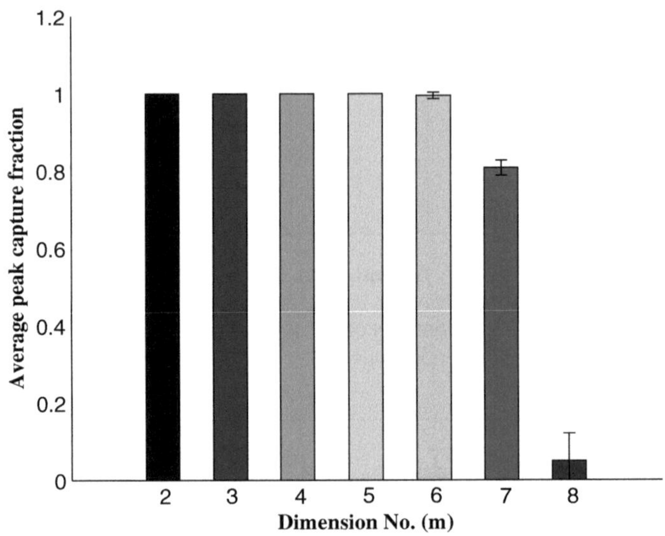

Fig. 11. Test results on Equal-peaks function: Plot of average fraction of peaks captured as a function of dimension number m (averaged over 30 trials)

edge or a corner of the search space making them difficult to be found by the algorithm. This observation can be supported by the results shown in Figures 9 and 10. Note that GSO performs well up to seven dimensions in the case of Rastrigin's function. However, good performance is found only up to $m = 5$ in the case of Schwefel's function. The performance of GSO on the Equal-peaks function (Figure 11) is similar to that of the Rastrigin's function up to six dimensions. Note the similarity in the shape of the function profiles of J_1 and J_3. However, the considered search space for J_3 is larger compared to that of J_1. This could explain why GSO's performance on J_3 is different from that on J_1 in seven and eight dimensions.

6 Concluding Remarks

We have used glowworm swarm optimization to address the problem of searching higher dimensional spaces. Results reported from tests conducted on Rastrigin's function reveals that GSO can effectively locate a total of 128 multiple peaks in seven dimensions. In this work, experiments in high dimensions have been limited to simple multimodal functions. Further work in this direction involves applying GSO to identification of multiple clusters in practical high dimensional databases. Literature on searching for multiple solutions in high dimensional spaces is rather meager and we hope that this work will introduce new ideas in this area.

References

Beasley, D., Bull, D.R., Martin, R.R.: A sequential niche technique for multimodal function optimization. Evolutionary Computation 1(2), 101–125 (1993)

Brits, R., Engelbrecht, A.P., van den Bergh, F.: A niching particle swarm optimizer. In: Proceedings of the 4th Asia-Pacific Conference on Simulated Evolution and Learning, pp. 692–696 (2002)

Clerc. Particle Swarm Optimization. ISTE Ltd., London (2007)

Goldberg, D., Richardson, J.: Genetic algorithms with sharing for multi-modal function optimization. In: Genetic Algorithms and Their Applications: Proceedings of the Second International Conference on Genetic Algorithms, pp. 44–49 (1987)

Kennedy, J.: Stereotyping: improving particle swarm performance with cluster analysis. In: Proceedings of the Congress on Evolutionary Computation, pp. 1507–1512 (2000)

Krishnanand, K.N.: Glowworm swarm optimization: a multimodal function optimization paradigm with applications to multiple signal source localization tasks. PhD thesis, Department of Aerospace Engineering, Indian Institute of Science (2007)

Krishnanand, K.N., Ghose, D.: Glowworm swarm based optimization algorithm for multimodal functions with collective robotics applications. Multiagent and Grid Systems 2(3), 209–222 (2006)

Krishnanand, K.N., Ghose, D.: Theoretical foundations for rendezvous of glowworm-inspired agent swarms at multiple locations. Robotics and Autonomous Systems 56(7), 549–569 (2008)

Krishnanand, K.N., Ghose, D.: Glowworm swarm optimization for simultaneous capture of multiple local optima of multimodal functions. Swarm Intelligence 3(2), 87–124 (2009a)

Krishnanand, K.N., Ghose, D.: Glowworm swarm optimisation: a new method for optimising multi-modal functions. Int. J. Computational Intelligence Studies 1(1), 93–119 (2009b)

Li, X.: Adaptively choosing neighbourhood bests using species in a particle swarm optimizer for multimodal function optimization. In: Deb, K., et al. (eds.) GECCO 2004. LNCS, vol. 3102, pp. 105–116. Springer, Heidelberg (2004)

Mühlenbein, H., Schomisch, D., Born, J.: The parallel genetic algorithm as function optimizer. Parallel Computing 17(6-7), 619–632 (1991)

Parsopoulos, K., Vrahatis, M.N.: On the computation of all global minimizers through particle swarm optimization. IEEE Transactions on Evolutionary Computation 8(3), 211–224 (2004)

Singh, G., Deb, K.: Comparison of multi-modal optimization algorithms based on evolutionary algorithms. In: Proceedings of the Genetic and Evolutionary Computation Conference, pp. 1305–1312 (2006)

Suganthan, P.N., Hansen, N., Liang, J.J., Deb, K., Chen, Y.P., Auger, A., Tiwari, S.: Problem defintions and evaluation criteria for the cec 2005 special session on real-parameter optimization. In: Technical Report, Nanyang Technological University, Singapore and KanGAL Report No. 2005005, IIT Kanpur, India (2005)

Törn, A., Zilinskas, A.: Global optimization. Springer, New York (1989)

5

Agent Specialization in Complex Social Swarms

Denton Cockburn and Ziad Kobti

School of Computer Science, University of Windsor, Windsor, Ontario, Canada

Abstract. The hypothesis that social influence leads to an increase in the division of labor, or specialization, in complex agent systems is introduced. Specialization, in turn, leads to increased productivity in such social systems. In this study, we examine the effect of social influence on the level of agent specialization in complex systems connected via social networks. Several methods attempt to explain the overall makeup of social influence and the emergence of specialization in general, with the most prominent being the genetic threshold model. This model posits that agents possess an inherent threshold for task stimulus, and when that threshold is exceeded, the agent will perform that task. The implication of social influence is that an agent's choice of which task to specialize in when multiple ones are available is influenced by the choices of its neighbours. Using the threshold model and an established metric that quantifies the level of agent specialization, we find that social influence indeed leads to an increase in the division of labour.

We further investigate the sensitivity of the social influence rate on the overall level of system specialization. The social influence rate is important in the social context because it determines how swayed an agent is by the decisions of other agents in its group. On one hand, a high social influence rate will cause agents to mimic the collective behaviour of its neighbours. On the other hand, a rate that is too small will cause agents to choose primarily based on genetic factors. Experimental results, by way of comparing different rate selection strategies, reveal that increases in the social influence rate causes increases in the agent specialization within the system.

1 Introduction

Agents are typically autonomous objects able to reason about their environment to maximize individual goals. Agent populations with communication and sophisticated social structure, likes that observed in ants, often yields to limitations in maximizing individual goal and instead settle for a goal appeasing to the population. An effect of socialization is specialization, that can be described as a subset of choices that an agent selects or frequents more than others. Social exchange and the distribution of specializations makes up for the lost individual productivity and increases reliance on the social network.

C.P. Lim et al. (Eds.): Innovations in Swarm Intelligence, SCI 248, pp. 77–89.
springerlink.com

The study of specialization is important to several fields. Social scientists and biologists for instance study specialization to better understand the changes in societies as a result of the emergence of specialization. It also gives insight into why individuals would choose to produce certain goods over others. From the biological perspective, specialization helps explain the behaviour of biological creatures such as ants, wasps and bees[10,16,15], which have been empirically shown to specialize based on tasks. Economically, specialization is studied to understand its effect upon a society's economy. It further serves to study how a market may grow or contract based on the specializations present, as these specializations lead to increases in the productivity of market systems [13]. Allyn Abbot Young points out that a productive individual increases the supply of certain commodities, while simultaneously increasing the demand for others [21]. This increase in productivity in turn increases per capita income according to Adam Smith [11]. Many of these fields use computer simulations to study the effect of specialization, so it is surprising that the simulation of specialization has not been studied more from a purely computer science perspective, with the exception of works such as [10].

There are several models that posit how specialization, also called division of labour, happens in groups. One common method is the genetic threshold model, which states that agents possess an inherent threshold for task stimulus, and when that threshold is exceeded, the agent will perform that task. This model is able to explain how caste systems evolve. We believe that there was also a social aspect to the idea of caste systems. For instance, we believe that having a lot of teachers in one's family will increase the likelihood of becoming a teacher. We hypothesize that such a practice would increase the level of specialization in a system, which in turn should lead to increased system productivity.

There is also a social aspect to division of labour, such as the activator/inhibitor studies [3,17]. These show that in certain insect colonies such as bees, certain bees evoke pheromones that suppress the desire of other agents to want to perform a task. Agents' desire to perform certain tasks would change as they age. This activator/inhibitor action would result in highly specialized colonies along the lines of age, in an effect called age polyethism. The idea of social influence is that an agent's choice of which task to specialize in when multiple ones are available, is influenced by the choices of its neighbours. Using a metric that measures the level of specialization within a system [7], we find that social influence leads to an increase in division of labour.

Social networks are an integral component of social systems. In this study we limit our networks to those exhibiting small-world properties, as these have been shown to resemble real world networks [2]. Some real-world networks that have been found to be small-world networks are the human acquaintance network and the world wide web. [12,5]. Keeping the focus on human agent specialization, we assume in this study that agents are autonomous, among other things, they are able to make their decisions based on probabilities, as opposed to simply following a rule function.

Many real-world social networks, such as the human acquaintance network, have been shown to be small-world networks, which are also explained below. We investigate the result of introducing social influence into small-world social networks, whereby an agent's choice of specialization is influenced by the choices of its neighbours. Given a choice between multiple specializations, we factor in what our neighbours are doing and allow that to influence our decision. We are limiting our study to autonomous agents, who make their decisions based on probabilities, as opposed to simply following a rule function. In addition, our agents' behaviours are also modeled after human agents, which fits into our research regarding the Pueblo people of the Mesa Verde region [9].

The object of this paper is to examine the effects of social influence on autonomous yet social swarms. Particularly sensitize the ranges of influence parameters on the degree of specialization emerging in such populations. In section 2 we describe the genetic threshold model and underlying generic mechanisms for task specialization. In section 3 we briefly introduce an overview of relevant definitions and concepts on small-world networks. Section 4 outlines the approach taken to model social influence. Experimental work and results are discussed in section 5. Finally, we conclude and summarize our finding with some highlights of potential future ventures.

2 Genetic Threshold Model

A number of variables contribute to the overall makeup of social influence and specialization in general. Included are genetic threshold level, economic demand level, and social influence rate. While very little attention has been paid to the underlying genetic mechanisms for task specialization [20], several genetic models have been put forward for the study of specialization. The most widely used is the response threshold model. The threshold model presents a certain level of stimulus for each task at which an individual will choose to specialize in that task [19]. The threshold model has been backed up by empirical evidence in insect societies such as ants and bees [15,6]. The response threshold model may in fact be an evolutionary behaviour, as agents that respond to problems (stimuli) quicker may be more likely to survive [8]. In the threshold model, agents by default perform no tasks. That is to say that if there is no stimulus for any of the possible tasks, then individuals will do nothing [3].

The determination of the level of stimulus present is dependent upon the context being studied. Even this has been approached from different angles, such as when all agents have the same stimulus levels for tasks, or when agents have differing thresholds based on individual genetic predispositions. In the latter case, low threshold individuals perform tasks at lower stimulus levels than others, and are thus more likely to specialize in those low stimulus tasks. The variance between individuals' threshold levels does not have to be significant, but seems necessary for the emergence of division of labour (DOL) in threshold models [8]. In some approaches, performing a task causes the threshold level for that task to decrease, while not performing the task will lead to the threshold level increasing

[19]. Most threshold models assume these thresholds to be fixed throughout the individual's lifetime or simply as a result of having performed or rejected the task previously, but empirical evidence suggests that threshold levels change over time in natural systems [3,16].

3 Small-World Networks

A social network is a social structure made up of related items. These networks are graph structures where connecting edges represent a relation between the two items. For example, a family tree is a social network, where each edge represents that the two connected nodes are relatives. The small-world phenomenon demonstrates that all humans, and agents in similar networks, are linked via short paths of acquaintances. This phenomenon has been observed in many networks that were not designed to exhibit such features. This includes the famous "6-degrees of separation" feature found within the US population [12] and similar features found concerning the World-Wide Web [5]. Small-world networks [12] are social networks that exhibit this phenomenon.

There are multiple ways to create a small-world network. A small-world network can be created using the Barabási-Albert model [1], which creates a scale-free network. Scale-free networks are networks whose degree distribution follows a power law pattern. This pattern is very common in real-world networks, and is the cause of the '6-degrees of separation' feature. In these power law distributed networks, nodes are not evenly connected. In large networks, there are many well-connected hubs and many sparsely connected nodes. The general ratio between these hubs and others remains constant even as the network's size changes [2].

A small-world network can also be created randomly. In these networks, nodes will tend to have a small, average amount of connections. Thus, the likelihood of hub nodes would be very small. These random networks therefore end up being a poor reflection of real-world social networks.

4 Approach Using Social Influence

Social influence is the concept that an agent's choice of specialization is influenced by the choices of those within its social network. Given a choice between multiple specializations, we factor in what the agent's neighbours are doing and allow that to influence the agent's decision. An agent N is defined as a neighbour of an agent A if N and A are directly connected within the social network. It has been shown that in the majority of cases, social influence causes an increase in the level of agent specialization when compared to systems with no social influence.

Waibel and his team in [20] also used the concept of social influence, but with what essentially were fully-connected networks (where all agents are connected to all others). In that study, agents chose their specialization based on the result of a function that weighs both their genetic threshold levels and inverse social influence. Therefore, chance of an agent choosing a task would reduce when more agents are already performing that task. This is similar to the activator/inhibitor model, but without the age factor implied in that model.

Our testing methodology is very similar to [8]. Both papers aim to measure changes in the level of division of labor. We used the Repast Agent Simulation Model [14].

We use the genetic threshold model among our agents. For each simulation run, each agent is assigned a threshold for each task. The average threshold for each task is 50, which is the same as in [8] and arbitrarily chosen. Since research has shown that the amount of threshold level variation (as long as some exists) between agents is irrelevant [8], we chose to use only one level. Each agent therefore only deviated in the range of ± 5 from the average threshold for each task. The agent's threshold for each task would differ, but would remain the same for the duration of each simulation.

Most studies involve 2-5 possible specializations [20]. Some insect colonies have anywhere from 20 to 40 specializations [3], while in a human environment, such as New York City, there may really be thousands of possible job choices. [8] demonstrated that group size and task number play a role in the level of specialization within a system. We therefore tested using 2, 5, 10, 20, and 100 tasks. For group size, we used 2, 10, 20, 100, 500, 1000. That means that we had tests with 2 tasks and 2 individuals, tests with 2 tasks and 10 individuals, and so on. We felt that these numbers would provide a clear impression of specialization under varying parameters.

All agents initially select a task at the beginning of the simulation run, with none of them being inactive. At each iteration, an agent has a chance γ of switching specialization. γ is the same as used in [8], and arbitrarily chosen to be 0.2. If no task meets the agent's threshold then the agent becomes inactive. If an agent is inactive at the beginning of an iteration, it attempts to choose a specialization. First, the agent a filters all the possible tasks, creating a set F where F = $\{S > T_{\mathrm{at}}\}$, for all tasks t. It should be noted that the inactive state for agents is not considered a task. $\mathrm{S_t}$ is the amount of stimulus available for the task t. T_{at} refers to the agent's threshold level for task t. The agent then selects a task from F using the two methods being compared in this paper. In the standard genetic threshold model, an agent selects a random task from F. We propose selecting a task from F that is influenced by the agent's network neighbours.

Given the set F, an agent selects one task, with a task t$\in$ F having probability: $\frac{1+\psi N_{\mathrm{at}}}{\sum(\psi N_{\mathrm{aj}})+\#F}$, for all tasks j in F, where $\mathrm{N_{at}}$ is the number of neighbours of agent a that are currently engaged in task t, and #F represents the total number of tasks in F. We use the symbol $\psi$to represent the influence impact an agent has on its neighbours selection. We will examine the sensitivity of the specialization level to changes in ψ by comparing results using varying levels. We use $\psi = 0.5$ as a baseline by which we compare other results. When selecting tasks that neighbours are engaged in, we ignore those neighbours that are currently inactive.

We tested with constant levels of ψ for all agents at 0, 0.2, 0.3, 0.4, 0.5, 0.6, 0.7, 0.8, 0.9 and 1. Note that when ψ is constant, all agents share the same influence rate. We also tested ψ with rates that vary between agents. For each level, we created a normal distribution with a set mean. These levels were 0.2$\pm$0.1, 0.3$\pm$0.1, 0.4$\pm$0.1, 0.5$\pm$0.1, 0.6$\pm$0.1, 0.7$\pm$0.1, 0.8$\pm$0.1, 0.9$\pm$0.1, 1$\pm$0.1, and 0.5$\pm$0.5.

The simulations also included the concept of demand. Demand is the total amount of effort needed to satisfy all tasks relative to the total work ability of all agents. A demand level $\delta < 1$ indicates that there is less work available than can be performed by all agents, thus potentially leading to inactive agents. We tested with demand levels of 0.7, 1, and 1.3.

At the beginning of each iteration, including the first, the stimulus level for each task is updated. The stimulus update formula is the same as used in [8]. Each agent performs the amount of work, a, which is arbitrarily assigned as 3. Each task Tj is updated by Bj $= a\frac{N}{T}\delta$, where N is the number of agents, T is the number of tasks, and δ is the previously mentioned demand level. This means that the stimulus level for a task is reduced when an agent performs that task. Therefore it is possible to exhaust the demand for each task, especially when the demand level δ is below 1.

At the beginning of each simulation, we create a social network using the Barabási-Albert [1] model. This model creates a small-world network, wherein all agents are connected to all other agents, but not necessarily directly. The connections between agents are fixed for the duration of each experiment. To compare the performance of the two strategies, we ensure that the created network is the same for both strategies. This does not mean that we use the same social network for every simulation. Each social network is used twice for each combination of parameters, once for the strategy using random selection model, and once using our social influence selection model, which is what we compare. In addition, each combination of parameters was tested with 10 different initial social networks, for a total of 1200 comparisons.

We use a method developed by Gorelick et al. to measure and compare our results. The method calculates and quantifies the degree to which agents are specialized [7]. It requires that the chosen specialization of all agents is recorded. We do this by having each active agent record its specialization at the end of each iteration. The recorded information of all agents is then stored in a nxm matrix, where n indicates each agent and m each task. The matrix is then normalized such that the sum of all cells is 1. The method developed in [7] then calculates the mutual information and Shannon entropy index [18]for the distribution of individuals across tasks. The result of dividing that mutual information score by the Shannon entropy score indicates how specialized agents were, with a result between 0 and 1. A score of 1 indicates that all agents are fully specialized, while 0 indicates no specialization. This is the same method used in [8]. More details of the methodology can be found in [7].

5 Results and Discussion

For uniform comparison of the levels of specialization, we first establish certain boundary tests by comparing levels of ($\psi = 0.5$) and ($\psi = 1 \pm 0.1$) to the standard genetic threshold model ($\psi = 0$) across varying parameters. Once we've established this, we proceed to use ($\psi = 0.5$) as our baseline to study the results at other levels. We normalize the results of each strategy by dividing the resulting

Table 1. DOL ratios at 0.7 demand level, with $\psi = 0.5$. Rows represent agent counts and columns represent task counts. Normalized based on performance of standard genetic threshold model ($\psi = 0$).

	2	10	50	100	500	1000
2	0	1.12	1.17	1.11	1.12	1.1
5	1.12	0.99	1	1.05	1.05	1.05
10	0.96	0.97	0.99	1.02	1.01	1.01
20	0.76	0.88	1.01	0.99	1	1
100	0.92	1.01	1	1	1	1

Table 2. DOL ratios at 0.9 demand level, with $\psi = 0.5$. Rows represent agent counts and columns represent task counts. Normalized based on performance of standard genetic threshold model ($\psi = 0$).

	2	10	50	100	500	1000
2	0	1.65	1.21	1.18	1.16	1.16
5	2.19	1	1.04	1.04	1.14	1.14
10	0.63	0.98	1.04	1.04	1.06	1.07
20	0.52	0.95	1	0.99	1.01	1.02
100	0.83	0.98	0.97	0.99	1	1

Table 3. DOL ratios at a demand level of 1, with $\psi = 0.5$. Rows represent agent counts and columns represent task counts. Normalized based on performance of standard genetic threshold model ($\psi = 0$).

	2	10	50	100	500	1000
2	0	1.09	1.34	1.32	1.2	1.23
5	1.9	1.08	1.03	1.11	1.2	1.2
10	0.65	0.99	1.1	1	1.18	1.17
20	0.54	1.02	1.06	0.83	1.09	1.1
100	0.77	1	1.04	0.86	1	1

of social influence by a baseline of ($\psi = 0$) in Tables 1-8 and ($\psi = 0.5$) in Tables 9-12. Therefore, a division of labour (DOL) ratio of 1.1 would indicate that there was 10% more average specialization than in the respective baseline model.

5.1 Comparison to Genetic Threshold Model ($\psi = 0$)

Notice that social influence increases the level of specialization, shown by cell numbers greater than 1. Even when demand level is lowest at 0.7, when social influence is expected to be lowest according to our tested parameters, there is still a general increase of specialization. Also notice that the effect of social influence also increases as the number of agents increase. Finally, it's visible that the effect

Table 4. DOL ratios at 1.3 demand level, with $\psi = 0.5$. Rows represent agent counts and columns represent task counts. Normalized based on performance of standard genetic threshold model ($\psi = 0$).

	2	10	50	100	500	1000
2	0	1.33	1.08	1.13	1.17	1.26
5	1.64	1.11	1.16	1.17	1.22	1.22
10	0.55	1.17	1.18	1.19	1.24	1.23
20	0.65	1.03	1.18	1.18	1.23	1.23
100	0.77	0.99	1.08	1.1	1.17	1.18

Table 5. DOL ratios at 0.7 demand level, with $\psi = 1 \pm 0.1$. Rows represent agent counts and columns represent task counts. Normalized based on performance of standard genetic threshold model ($\psi = 0$).

	2	10	50	100	500	1000
2	7.96	1.09	1.1	1.05	1.11	1.16
5	1.86	1.12	0.97	1.01	1.05	1.04
10	1.01	1	1.01	1.01	1.01	1.02
20	0.92	1.08	0.99	1	1	1
100	0.86	0.99	1	1	1	1

Table 6. DOL ratios at 0.9 demand level, with $\psi = 1 \pm 0.1$. Rows represent agent counts and columns represent task counts. Normalized based on performance of standard genetic threshold model ($\psi = 0$).

	2	10	50	100	500	1000
2	4.04	1.71	1.05	1.3	1.18	1.19
5	1.48	1.17	1.12	1.1	1.15	1.15
10	0.97	0.99	1.01	1.03	1.07	1.08
20	1.06	1.09	1.01	1.01	1.02	1.02
100	0.92	1.01	1	1	1	1

Table 7. DOL ratios at demand level of 1, with $\psi = 1 \pm 0.1$. Rows represent agent counts and columns represent task counts. Normalized based on performance of standard genetic threshold model ($\psi = 0$).

	2	10	50	100	500	1000
2	14.21	1.69	1.32	1.13	1.3	1.22
5	0.67	1.37	1.36	1.15	1.22	1.25
10	1.19	1.09	1.09	1.09	1.18	1.2
20	0.95	0.97	1.09	1.02	1.1	1.1
100	0.90	0.99	1.05	1	1	1.01

Table 8. DOL ratios at 1.3 demand level, with $\psi = 1 \pm 0.1$. Rows represent agent counts and columns represent task counts. Normalized based on performance of standard genetic threshold model ($\psi = 0$).

	2	10	50	100	500	1000
2	17.26	1.63	1.53	1.38	1.27	1.29
5	0.69	1.03	1.22	1.25	1.32	1.29
10	1.05	1.08	1.13	1.24	1.27	1.28
20	0.76	1.05	1.15	1.22	1.24	1.26
100	0.85	1.05	1.06	1.1	1.15	1.18

also increases as the demand level increases. Note the general growth of the ratio as we move from a demand level of 0.7 through to a demand level of 1.3. This is in spite of the fact that division of labour is known to drop significantly (almost nil) as demand levels exceed 1. We set the value of the cell at 1x1 to 0, as the level of specialization was so low that it rendered any meaningful comparison meaningless.

When demand level is below 1, agents have fewer specializations that will have enough stimulus to surpass their thresholds. As lower stimulus lead to more inactive agents, the chance of these agents suddenly coming upon multiple tasks that were under-worked is slim. We expected little to no increase from social influence here. This is due to the fact that social influence only plays a role when an agent has multiple specializations from which to choose. The results indicate that even when there is low demand, enough agents are still faced with multiple choices, resulting in an specialization from social influence.

When the demand level is 1, the amount of stimulus added is exactly equal to the amount of work able to be performed. Excess stimulus only remains when agents are not properly tasked - either because they are inactive, or on a task which doesn't have enough work for them to perform. There will be a few agents that are misplaced (inactive or under-productive), resulting in a slight increase in the level of excess stimulus per iteration. This situation will lead to more agents having multiple tasks above their threshold level. This causes a further increase in specialization than that seen at the lower levels.

A demand level above 1 indicates that even if all agents are active and fully worked, they are not able to satisfy all the stimulus in the system. As the level of excess stimulus increases per iteration, agents will have more tasks available that surpass its threshold. When an agent's neighborhood remains consistent - where its neighbours are maintaining their specialization, there is a higher probability of it choosing the more popular tasks, as determined by its neighbours. We found that while specialization was very low when demand level was above 1, there was still an increase from social influence.

Because the choice that an agent makes is based on a probability, there still remains a chance that an agent will choose a specialization that none of its neighbours have chosen. As this will in turn influence those neighbours decisions, its possible that in a few cases social influence may actually decrease specialization over a short period of time. This is especially pronounced in small networks, as the effect of choosing new specializations will cascade more quickly, having a greater effect on the entire network. We believe this to be the explanation for the cases in our experiments where there is a reduced level of specialization where an increase is expected.

We also noticed that agents striving to "follow the Joneses" when choosing specializations also result in a sort of topology based caste system. In cases where demand level is significantly above 1, agents will tend more to choose the same specializations as their neighbours. This has a cascading effect, as one agent changing its specialization will influence its neighbours to also follow suit. In the extreme case where tasks have more stimulus than can be performed by all agents, the result will be that all agents may converge upon performing one task. Even when all agents have converged, there is still a chance that an agent may choose a specialization not performed by any of its neighbours, even if all agents are directly connected. We can deduce that the likelihood of convergence is also dependent upon the connectivity of the social network. The more neighbours an agent has, the more influence exerted upon its decisions.

5.2 Comparison to Fixed Rate of $\psi = 0.5$

When ψ is at a constant level below 0.5 performance is worse than our baseline ($\psi = 0.5$). This can be seen in Table 1. There is a pattern of increasing performance levels also visible in Table 1, with the exception being the small drop going from $\psi = 0.1$ to $\psi = 0.2$. This pattern is also evident in Table 2, where again there is an exception with a small drop going from $\psi = 0.9$ to $\psi = 1$. There is also a pattern of increasing levels of specialization while ψ continues past 0.5.

When we compare $\psi = 0.5$ with a varying rate of $\psi = 0.5 \pm 0.1$, we see that varying rates between agents produce higher performance, even when they still

Table 9. Result of other social influence rate strategies compared to fixed rate of $\psi = 0.5$ when ran over 1200 experiments

Specialization	$\psi = 0$	$\psi = 0.1$	$\psi = 0.2$	$\psi = 0.3$	$\psi = 0.4$
Increased	384 (32.0%)	404 (33.7%)	395 (32.9%)	430 (35.8%)	481 (40.1%)
No change	112 (9.3%)	2 (0.2%)	0	0	0
Decreased	704 (58.7%)	794 (66.2%)	805 (67.1%)	770 (64.2%)	719 (59.9%)

Table 10. Result of other social influence rate strategies compared to fixed rate of $\psi = 0.5$ when ran over 1200 experiments

Specialization	$\psi = 0.6$	$\psi = 0.7$	$\psi = 0.8$	$\psi = 0.9$	$\psi = 1.0$
Increased	718 (59.8%)	752 (62.7%)	785 (65.4%)	799 (66.6%)	796 (66.3%)
No change	0	0	0	0	0
Decreased	482 (40.2%)	448 (37.3%)	415 (34.6%)	401 (33.4%)	404 (33.7%)

Table 11. Result of other social influence rate strategies compared to fixed rate of ψ = 0.5 when ran over 1200 experiments

Specialization	$\psi = 0.2 \pm 0.1$	$\psi = 0.3 \pm 0.1$	$\psi = 0.4 \pm 0.1$	$\psi = 0.5 \pm 0.1$	$\psi = 0.6 \pm 0.1$
Increased	379 (31.6%)	362 (30.2%)	403 (33.6%)	645 (53.8%)	647 (53.9%)
No change	113 (9.4%)	137 (11.4%)	151 (12.6%)	174 (14.5%)	149 (12.4%)
Decreased	708 (59.0%)	701 (58.4%)	646 (53.8%)	381 (31.8%)	404 (33.7%)

Table 12. Result of other social influence rate strategies compared to fixed rate of ψ = 0.5 when ran over 1200 experiments

Specialization	$\psi = 0.7 \pm 0.1$	$\psi = 0.8 \pm 0.1$	$\psi = 0.9 \pm 0.1$	$\psi = 1.0 \pm 0.1$	$\psi = 0.5 \pm 0.5$
Increased	699 (58.3%)	743 (61.9%)	753 (62.8%)	761 (63.4%)	571 (47.6%)
No change	136 (11.3%)	120 (10.0%)	105 (8.8%)	109 (9.1%)	172 (14.3%)
Decreased	365 (30.4%)	337 (28.1%)	342 (28.5%)	330 (27.5%)	457 (38.1%)

average the same. Specialization levels also increase as the social influence rate increases when using varying rates. There is a drop going from $\psi = 0.2 \pm 0.1$ to $\psi = 0.3 \pm 0.1$, with this being the only exception. $\psi = 0.5 \pm 0.1$ produces higher levels of specialization than $\psi = 0.5 \pm 0.5$, which suggests that increasing the variance rate does not produce more specialization.

In our previous work, we demonstrated the increase of specialization levels by comparing results when $\psi = 0.5$ and $\psi = 0$ (genetic threshold model). We presented these results over a varying range of agent counts and task amounts. The values tested were the same values tested in our present experiments. These are presented again below in Tables 5-8. To highlight the increase in performance resulting from increasing ψ, we present the result of our comparisons of $\psi = 1 \pm 0.1$ with $\psi = 0$. These can be seen in Tables 9-12. We can see that $\psi = 1 \pm 0.1$ produces higher specialization than $\psi = 0$ in the vast majority of cases, even across different parameter settings. This further reinforces our finding that increasing the social influence rate will lead to increased specialization.

6 Conclusion and Future Work

In this study we examined the effects of a social influence strategy on an agent's choice of specialization. We found that social influence increases division of labour when there is an excess of demand. When influenced by their peers, agents become more specialized when there is too much task choice available. Our results reinforce the findings of previous research that indicate that specialization increases as group size and task number increases, which is also found in human societies [4]. Our research further shows that in such settings, social influence increases the level of specialization above and beyond the increase from those factors. It can also be concluded that increasing the influence rate lead to increases in the level of agent specialization. This was found to be true in cases where agents shared the same constant influence rate, and cases where agents had varying influence rates.

While it is possible still to increase ψ beyond a value of 1, we don't think it wise to do so. The higher the value of ψ, the higher the likelihood that an agent will choose a specialization from those chosen by its neighbours. Given a system with a large number of tasks and a small level of agent connectivity, agents are much less likely to choose tasks not being performed by others. While this will result in a higher level of specialization, the counterpoint is that there will be a reduced level of diversity. More agents will be performing the same tasks. If a specialization has an effect on system resources, then this may result in one resource becoming depleted rapidly. This is in spite of the fact that agents will only perform specializations while demand exists. Demand may exist for items that are in danger of depletion, such as oil is in the real world.

In the future, we would like to investigate the effect of social influence on stimulus perception. In this paper, the effect of social influence after a task has already surpassed the agent's threshold. We would like to study the idea that if more of our neighbours are performing a task, it becomes "cooler". The underlying idea being that such an effect may reduce an agent's innate threshold for that task, allowing them to perform tasks which they may not have if they were not influenced. Furthermore, it would be interesting to examine other social strategies for division of labour, perhaps with the goal of comparing these to find the most effective. One such idea is to investigate the desire of agents to find niches. We are interested in studying what effect the drive to be different has on division of labour. This was addressed in [20], where agents chose their specialization based on the result of a function that weighs both their threshold levels and inverse social influence. We believe that there is more room for investigation with regards to that approach. In keeping with that view, we would like to compare these different approaches, perhaps even hybridizing them, to understand the effect.

Acknowledgments

This work was partially funded by National Science Foundation and NSERC Discovery grants.

References

1. Barabási, A., Albert, R.: Emergence of scaling in random networks. Science 286, 509–512 (1999)
2. Barabási, A., Albert, R.: Scale-free networks. Scientific American 288, 60–69 (2003)
3. Beshers, S., Fewell, J.: Models of division of labor in social insects. Annu. Rev. Entomol. 46, 413–440 (2001)
4. Bonner, J.: Dividing the labor in cells and societies. Current Science 64, 459–466 (1993)
5. Bu, T., Towsley, D.: On distinguishing between internet power law topology generators. In: INFOCOM (2002)

6. Fewell, J., Page, R.: Colony-level selection effects on individual and colony foraging task performance in honeybees, apis mellifera l. Behav. Ecol. Sociobiology 48, 173–181 (2000)
7. Gorelick, R., Bertram, S.K.P.F.J.: Normalized mutual entropy in biology: quantifying division of labor. American Naturalist 164, 678–682 (2004)
8. Jeanson, R., Fewell, J., Gorelick, R., Bertram, S.: Emergence of increased division of labor as a function of group size. Behavioral Ecology and Sociobiology (2007)
9. Kobti, Z., Reynolds, R.K.T.: The emergence of social network hierarchy using cultural algorithms. International Journal on Artificial Intelligence Tools 15(6), 963–978 (2006)
10. Larsen, J.: Specialization and division of labour in distributed autonomous agents. Master's thesis, University of Aarhus (2001)
11. Lavezzi, A.M.: Smith, marshall and young on division of labour and economic growth. European Journal of the History of Economic Thought 10, 81–108 (2003)
12. Milgram, S.: The small world problem. Psychology Today, 60–67 (1967)
13. Murciano, A., Millán, J., Zamora, J.: Specialization in multi-agent systems through learning. Biological Cybernetics 76(5), 375–382 (1997)
14. North, M.J., Collier, N.V.J.: Experiences creating three implementations of the repast agent modeling toolkit. ACM Transactions on Modeling and Computer Simulation 16(1), 1–25 (2006)
15. O'Donnell, S.: Rapd markers suggest genotypic effects on forager specialization in a eusocial wasp. Behav. Ecol. Sociobiology 38, 83–88 (1996)
16. Page, R., Erber, J., Fondrk, M.: The effect of genotype on response thresholds to sucrose and foraging behavior of honey bees (apis mellifera l.). J. Comp. Physiol. A 182, 489–500 (1998)
17. Ravary, F., Lecoutey, E., Kaminski, G., Châline, N., Jaisson, P.: Individual experience alone can generate lasting division of labor in ants. Curr. Biol. 17, 1308–1312 (2007)
18. Shannon, C.: A mathematical theory of communication. Bell System Technical Journal 27, 379–423, 623–656 (1948)
19. Theraulaz, G., Bonabeau, E., Deneubourg, J.: Response threshold reinforcements and division of labour in insect societies. Proc. R. Soc. Lond. B. Biol. Sci. 265, 327–332 (1998)
20. Waibel, M., Floreano, D., Magnenat, S., Keller, L.: Division of labour and colony efficiency in social insects: effects of interactions between genetic archi- tecture, colony kin structure and rate of perturbations. R. Soc. Lond B (273), 1815–1823 (2006)
21. Young, A.A.: Increasing returns and economic progress. The Economic Journal 38, 527–542 (1928)

6

Computational Complexity of Ant Colony Optimization and Its Hybridization with Local Search

Frank Neumann[1], Dirk Sudholt[2,⋆], and Carsten Witt[3,⋆]

[1] Max-Planck-Institut für Informatik, 66123 Saarbrücken, Germany
[2] Informatik 2, Technische Universität Dortmund, 44221 Dortmund, Germany
[3] DTU Informatics, Technical University of Denmark, 2800 Kgs. Lyngby, Denmark

Abstract. The computational complexity of ant colony optimization (ACO) is a new and rapidly growing research area. The finite-time dynamics of ACO algorithms is assessed with mathematical rigor using bounds on the (expected) time until an ACO algorithm finds a global optimum. We review previous results in this area and introduce the reader into common analysis methods. These techniques are then applied to obtain bounds for different ACO algorithms on classes of pseudo-Boolean problems. The resulting runtime bounds are further used to clarify important design issues from a theoretical perspective. We deal with the question whether the current best-so-far solution should be replaced by new solutions with the same quality. Afterwards, we discuss the hybridization of ACO with local search and present examples where introducing local search leads to a tremendous speed-up and to a dramatic loss in performance, respectively.

1 Introduction

The fascinating collective behavior of swarms is a rich source of inspiration for the field of computational intelligence (see, e. g., Kennedy, Eberhart, and Shi, 2001). In particular, the ability of ants to find shortest paths has been transferred to shortest path problems in graphs (Dorigo and Stützle, 2004). The idea is that artificial ants traverse the graph from a start node (nest) to a target node (food). On each edge a certain amount of artificial pheromone is deposited. At each node each ant chooses which edge to take next. This choice is made probabilistically and according to the amount of pheromone placed on the edges. As in real ant colonies, the pheromones evaporate over time. The amount of evaporation is determined by the so-called *evaporation factor* ρ, $0 < \rho < 1$. In every pheromone update on every edge a ρ-fraction of the pheromone evaporates, i. e., if the edge contains pheromone τ, the remaining amount of pheromone is $(1-\rho)\cdot\tau$ and

⋆ This author was supported by the Deutsche Forschungsgemeinschaft (DFG) as a part of the Collaborative Research Center "Computational Intelligence" (SFB 531).

C.P. Lim et al. (Eds.): Innovations in Swarm Intelligence, SCI 248, pp. 91–120.
springerlink.com

then eventually new pheromone is added. Intuitively, a large evaporation factor implies that the impact of previously laid pheromones diminishes quickly and new pheromones have a large impact on the system. Small evaporation factors, on the other hand, imply that the system only adapts slowly to new pheromones.

In contrast to real ants, however, new pheromones are often not placed immediately after traversing an edge. In order to avoid rewarding cycles or paths leading to dead ends, pheromones are usually placed after the ant has found the target node and only for edges that are not part of a cycle on the ant's trail. Also, in such an artificial system the amount of pheromone placed may depend on the length of the constructed path, so that short paths are rewarded more than longer paths. Such an adaptation is often necessary as the movement of artificial ants is usually synchronized, in contrast to real ants in nature, where the order of ants arriving at a location is essential.

Optimization with artificial ants is known as *ant colony optimization* (ACO). ACO algorithms have been used for shortest path problems, but also for the well-known Traveling Salesman Problem (TSP) and routing problems (Dorigo and Stützle, 2004). However, the use of artificial ants is not limited to graph problems. ACO algorithms can also be used to construct solutions for combinatorial problems, e. g., for pseudo-Boolean functions. In the graph-based ant system (GBAS) by Gutjahr (2000) a solution is constructed by letting an artificial ant traverse a so-called *construction graph* and mapping the path chosen by the ant to a solution for the problem.

The popularity of ACO algorithms has grown rapidly in recent years. The research effort spent to develop, design, analyze, and apply ACO algorithms is demonstrated by a huge number of applications and publications. In order to understand the working principles of ACO algorithms and their dynamic behavior, a solid theoretical foundation is needed. Until 2006 only convergence results were known for ACO algorithms (Gutjahr, 2003) or investigations of simplified models of ACO algorithms (Merkle and Middendorf, 2002). Results on simplified models can give good hints about the real algorithm without simplifying assumptions. However, without bounds on the errors introduced in the simplifications it is usually not possible to draw rigorous conclusions for the real algorithm. Convergence results state that as the time goes to infinity the probability of finding the optimum tends to 1. However, these results do not yield insights into the finite-time behavior of ACO algorithms.

In this chapter we review a new line of research in ACO theory: the computational complexity or runtime analysis of ACO algorithms. The finite-time behavior of ACO algorithms is analyzed using (asymptotic) bounds on the expected time an ACO algorithm needs to find an optimum. These bounds are obtained with mathematical rigor using tools from the analysis of randomized algorithms and they can be used to predict and to judge the performance of ACO algorithms for arbitrary problem sizes.

The analysis of the computational complexity of randomized search heuristics has already been follows for evolutionary algorithms, with great success. The analysis of evolutionary algorithms started for simply structured example functions with interesting properties (see e. g. Droste, Jansen, and Wegener (2002)). Those toy problems encouraged or demanded the development of new methods and tools and revealed important insights into the working principles of evolutionary algorithms. This then allowed the analysis of more sophisticated artificial problems and, finally, the analysis of problems from combinatorial optimization (see e. g. Giel and Wegener (2003); Neumann and Wegener (2007); Witt (2005)).

It therefore makes sense to start the investigation of ACO algorithms with simply structured example functions as well. This approach has been explicitly demanded in a survey by Dorigo and Blum (2005) for simple pseudo-Boolean problems. In pseudo-Boolean optimization the goal is to maximize a function $f\colon \{0,1\}^n \to \mathbb{R}$. The following functions are well known from the analysis of evolutionary algorithms and have also been considered in the initial rigorous runtime analyses of ACO algorithms.

$$\begin{aligned} \textsc{OneMax}(x) &= \sum_{i=1}^{n} x_i \\ \textsc{LeadingOnes}(x) &= \sum_{i=1}^{n} \prod_{j=1}^{i} x_j \\ \textsc{BinVal}(x) &= \sum_{i=1}^{n} 2^{n-i} x_i \end{aligned}$$

The function OneMax simply counts the number of bits set to 1. This function may also be rediscovered in a practical problem when a specific target point has to be hit and the objective function gives best possible hints towards this point. LeadingOnes counts the number of leading ones in the bit string and BinVal equals the binary value of the bit string. For these problems the bits have different priorities. For BinVal some local changes have a stronger effect than others. In LeadingOnes the bits to the right do not affect the function value until the bits to the left have been set correctly.

The runtime analysis of ACO algorithms has been started independently by Gutjahr (2008) and Neumann and Witt (2006) for (variations of) the function OneMax. Gutjahr (2008) analyzed a GBAS algorithm with a chain construction graph using a slightly different pheromone update rule on a class of generalized OneMax functions. For a fairly large value of ρ, he proved an upper bound of $O(n \log n)$ on the expected number of function evaluations. Neumann and Witt (2006) independently studied a simple ACO algorithm called 1-ANT. The 1-ANT keeps track of the best ant solution found so far. If an iteration creates a solution that is not worse than the best-so-far solution, a pheromone update with respect to the new solution is performed. Otherwise, all pheromones remain unchanged. Another way of putting it is that every new best-so-far solution is reinforced only once. Interestingly, Neumann and Witt (2006) proved that on the function OneMax the evaporation factor ρ has a tremendous impact on the performance of the 1-ANT. If ρ is larger than a certain threshold, the 1-ANT behaves similar to the (1+1) EA. On the other hand, if ρ is below the threshold, the 1-ANT tends to stagnate. As the best-so-far OneMax-value increases more and more, the 1-ANT is forced to discover solutions with more and more 1-bits. However, the impact of the following pheromone update is so small that 1-bits are not rewarded enough and even rediscovering previously set 1-bits gets increasingly harder. This leads to stagnation and exponential runtimes. A similar behavior was proven by Doerr, Neumann, Sudholt, and Witt (2007) for the functions LeadingOnes and BinVal. On these functions the phase transition between polynomial and superpolynomial optimization times could be identified more precisely with asymptotically tight bounds. Beside the mentioned results on pseudo-Boolean example functions, there are also results for some classical combinatorial optimization problems. Neumann and Witt (2008) investigated the impact of the construction graph in ACO algorithms for the minimum spanning tree problem and Attiratanasunthron and Fakcharoenphol (2008) analyzed an ACO algorithm for the shortest path problem on directed acyclic graphs.

A lesson learned from the analysis of the 1-ANT is that performing an update only in case a new best-so-far solution is found may, in general, not be a good design choice. In fact, many ACO algorithms used in applications use *best-so-far reinforcement* where in every iteration the current best-so-far solution is reinforced. In other words, in every iteration a pheromone update happens, using either the old or a newly generated best-so-far solution. We show how these algorithms, variants of the MAX-MIN ant system (MMAS) (see Stützle and Hoos, 2000), can be analyzed for various example functions, including the class of unimodal functions (Section 3) and plateau functions (Section 4). Thereby, we follow the presentation from Neumann, Sudholt, and Witt (2009), which extends previous work by Gutjahr and Sebastiani (2008). The latter authors first analyzed MMAS variants on OneMax, LeadingOnes, and functions with plateaus. Their results and our contributions show that the impact of ρ is by far not as drastic as for the 1-ANT. When decreasing ρ, the algorithms become more and more similar to random search and the runtime on simple functions grows with $1/\rho$, but there is no phase transition for polynomially small ρ as for the 1-ANT. We also demonstrate how (a restricted formulation of) the fitness-level method can be adapted to the analysis of ACO algorithms. Finally, we present lower bounds for ACO algorithms: a general lower bound for functions with unique optimum that grows with $1/\rho$ and an almost tight lower bound for LeadingOnes.

Often evolutionary algorithms or ACO algorithms are combined with local search procedures. The effect of combining evolutionary algorithms with local search methods has already been examined by Sudholt (2008, 2009). In a typical ACO algorithm without local search, old and new best-so-far solutions are typically quite close to one another and the distribution of constructed ant solutions follows the most recent best-so-far solutions. When introducing local search, old and new best-so-far solutions might be far apart. We discuss this effect in more detail in Section 5, including possible effects on search dynamics. To exemplify the impact on performance, we present an artificial function where an ant algorithm with local search drastically outperforms its variant without local search and a second function where the effect is reversed.

2 Basic Algorithms

In the following investigations we restrict ourselves to pseudo-Boolean optimization. The goal is to maximize a function $f\colon \{0,1\}^n \to \mathbb{R}$, i.e., the search space consists of bit strings of length n. These bit strings are synonymously called solutions or search points.

A natural construction graph for pseudo-Boolean optimization is known in the literature as *Chain* (Gutjahr, 2008). We use a simpler variant as described by Doerr and Johannsen (2007), based on a directed multigraph $C = (V, E)$. In addition to a start node v_0, there is a node v_i for every bit i, $1 \le i \le n$. This node can be reached from v_{i-1} by two edges. The edge $e_{i,1}$ corresponds to setting bit i to 1, while $e_{i,0}$ corresponds to setting bit i to 0. The former edge is also called a *1-edge*, the latter is called *0-edge*. An example of a construction graph for $n = 5$ is shown in Figure 1.

In a solution construction process an artificial ant sequentially traverses the nodes $v_0, v_1, \ldots, v_n$. The decision which edge to take is made according to pheromones on the edges. Formally, we denote pheromones by a function $\tau\colon E \to \mathbb{R}_0^+$. From v_{i-1} the edge $e_{i,1}$ is taken with probability $\tau(e_{i,1})/(\tau(e_{i,0}) + \tau(e_{i,1}))$. We give a formal description of this procedure for arbitrary construction graphs that returns a path P taken by the ant. In the case of our construction graph, we also identify P with a binary solution x as described above and denote the path by $P(x)$.

All ACO algorithms considered in this chapter start with an equal amount of pheromone on all edges: $\tau(e_{i,0}) = \tau(e_{i,1}) = 1/2$. Moreover, we ensure that $\tau(e_{i,0}) + \tau(e_{i,1}) = 1$ holds, i.e., pheromones for one bit always sum up to 1. This implies that the probability of taking a specific edge equals its pheromone value; in other words, pheromones and traversion probabilities coincide. We remark that the solution construction in ACO algorithms is often enhanced by incorporating heuristic information for parts of the solution. The probability of choosing an edge then depends on both the pheromones and the heuristic information.

Given a solution x and a path $P(x)$ of edges that have been chosen in the creation of x, a pheromone update with respect to x is performed as follows. First, a ρ-fraction of all pheromones evaporates and a $(1-\rho)$-fraction remains. Next, some pheromone is added to edges that are part of the path $P(x)$ of x. However, with these simple rules we cannot exclude that pheromone on some edges may converge to 0, so that the probability of choosing this edge becomes

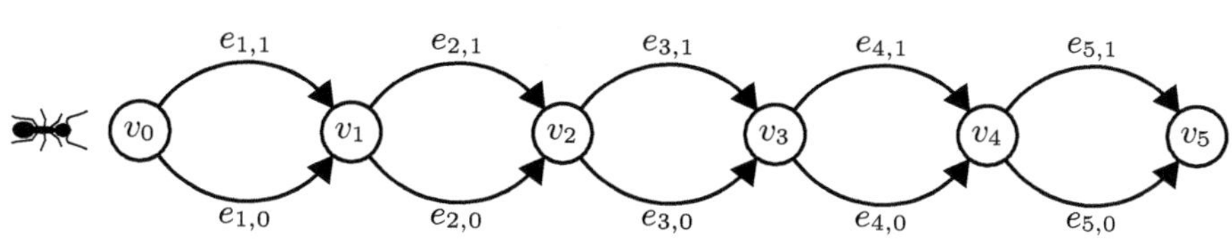

Fig. 1. Construction graph for pseudo-Boolean optimization with $n = 5$ bits.

nearly 0. In such a setting, the algorithm has practically no chance to revert a wrong decision. To prevent pheromones from dropping to arbitrarily small values, we follow the MAX-MIN ant system by Stützle and Hoos (2000) and restrict all pheromones to a bounded interval. The precise interval is chosen as $[1/n, 1-1/n]$. This choice is inspired by standard mutations in evolutionary computation where for every bit an evolutionary algorithm has a probability of $1/n$ of reverting a wrong decision.

Operator 1. Construct(C, τ)

Let $P := \emptyset$, $v := v_0$, and mark v as visited.
repeat
  Let E_v be the set of edges leading to non-visited successors of v in C.
  if $E_v \neq \emptyset$ **then**
    Choose $e \in E_v$ with probability $\tau(e)/\sum_{e'|e' \in E_v} \tau(e')$.
    Let $e = (v, w)$, mark w as visited, set $v := w$, and append e to P.
until $E_v = \emptyset$.
return the constructed path P.

Depending on whether an edge e is contained in the path $P(x)$ of the solution x, the pheromone values τ are updated to τ' as follows:

$$\begin{aligned} \tau'(e) &= \min\left\{(1-\rho)\cdot\tau(e)+\rho,\ 1-\frac{1}{n}\right\} && \text{if } e \in P(x) \text{ and} \\ \tau'(e) &= \max\left\{(1-\rho)\cdot\tau(e),\ \frac{1}{n}\right\} && \text{if } e \notin P(x). \end{aligned} \tag{1}$$

Note that the pheromones on all 1-edges suffices to describe all pheromones. They form a probabilistic model that somehow reflects a collective memory of many previous paths found by artificial ants.

We consider the runtime behavior of several ACO algorithms that all use the above construction procedure to construct new solutions. One such algorithm is called MMAS_{bs} in Gutjahr and Sebastiani (2008); we refer to it as MMAS*. It is shown on the right-hand side in Figure 2. Note that MMAS* only accepts strict improvements when updating the best-so-far solution. We also investigate a variant of MMAS_{bs} that accepts solutions of equal quality as this enables the algorithm to explore plateaus, i. e., regions of the search space with equal function value. We call this algorithm MMAS, it is displayed on the left-hand side of Figure 2. In Section 3, we will analyze the runtime behavior of MMAS* and MMAS on simple unimodal functions.

ACO algorithms like MMAS* are not far away from evolutionary algorithms. If the value of ρ is chosen large enough in MMAS*, the pheromone borders $1/n$ or $1-1/n$ are touched for every bit. In this case, MMAS* becomes the same as the algorithm called (1+1) EA*, which is known from the analysis of evolutionary algorithms (Jansen and Wegener, 2001).

MMAS

Set $\tau(e) := 1/2$ for all $e \in E$.
Construct a solution x^*.
Update pheromones w. r. t. x^*.
repeat
  Construct a solution x.
  if $f(x) \geq f(x^*)$ **then** $x^* := x$.
  Update pheromones w. r. t. x^*.

MMAS*

Set $\tau(e) := 1/2$ for all $e \in E$.
Construct a solution x^*.
Update pheromones w. r. t. x^*.
repeat
  Construct a solution x.
  if $f(x) > f(x^*)$ **then** $x^* := x$.
  Update pheromones w. r. t. x^*.

(1+1) EA

Choose x^* uniformly at random.
repeat
  Create x by flipping each bit in x^*
    independently with prob. $1/n$.
  if $f(x) \geq f(x^*)$ **then** $x^* := x$.

(1+1) EA*

Choose x^* uniformly at random.
repeat
  Create x by flipping each bit in x^*
    independently with prob. $1/n$.
  if $f(x) > f(x^*)$ **then** $x^* := x$.

Fig. 2. The algorithms considered in Section 3 and 4. The starred variants on the right-hand side use a strict selection. The left-hand side algorithms also accept solutions of equal quality. If ρ is so large that the pheromone borders are hit in a single pheromone update, MMAS collapses to the (1+1) EA and MMAS* collapses to the (1+1) EA*.

As already pointed out by Jansen and Wegener (2001), the (1+1) EA* has difficulties with simple plateaus of equal quality as no search points of the same function value as the best so far are accepted. Accepting solutions with equal quality enables the algorithm to explore plateaus by random walks. Therefore, it seems more natural to replace search points by new solutions that are at least as good. In the case of evolutionary algorithms, this leads to the (1+1) EA which differs from the (1+1) EA* only in the acceptance criterion. For the sake of completeness, both the (1+1) EA* and the (1+1) EA are also shown in Figure 2. By essentially the same arguments, we also expect MMAS to outperform MMAS* on plateau functions. The corresponding runtime analyses are presented in Section 4. Finally, Section 5 deals with a hybrid of MMAS* and local search. This algorithm is defined in Section 5.

3 On the Analysis of ACO Algorithms

We are interested in the random optimization time or runtime, that is, the number of iterations until a global optimum is sampled first. As both MMAS* and MMAS only evaluate a single new solution in each iteration, this measure equals the number of function evaluations for these algorithms. Often the expected optimization time is considered and we are interested in bounding this expectation from above and/or below. The resulting bounds are asymptotic and stated using the common notation for asymptotics (see, e. g., Cormen, Leiserson, Rivest, and Stein, 2001). Note that in evolutionary computation an iteration is commonly called a *generation* and the function value is called *fitness*.

The analysis of simple ACO algorithms is more complicated than the analysis of evolutionary algorithms. Typical evolutionary and genetic algorithms can be modelled as Markov chains as the next generation's population only depends on the individuals in the current population. In ACO pheromone traces evaporate only slowly, hence their impact on pheromones is present for a much longer period of time. This means that the next iteration's state typically depends on many previous iterations.

In addition, the state space for the probabilistic model is larger than for evolutionary algorithms. For the simple (1+1) EA there is for each bit a probability of either $1/n$ or $1-1/n$ of creating a 1 in the next solution. For an ACO algorithm this probability equals the pheromone on the corresponding 1-edge, which can attain almost arbitrary values in the interval $[1/n,\ 1-1/n]$ unless ρ is very large. Note, however, that the number of possible pheromone values for one bit is countably infinite as in every iteration at most two new values can be attained.

The probability of creating a specific ant solution equals the product of probabilities of creating the respective bit value for each bit. This holds since the construction procedure treats all bits independently. At initialization this probability is $(1/2)^n$ for each specific solution as all pheromones equal $1/2$. However, when search becomes more focused and all pheromones reach their borders, the most likely solution has a probability of $(1-1/n)^n \approx 1/e$, $e = \exp(1)$, of being (re-)created. In order to understand the dynamic behavior of ACO algorithms it is essential to understand the dynamic behavior of pheromones for specific bits.

Many pseudo-Boolean problems contain bits where a certain bit value is "good" in a sense that a good bit value leads to a better function value than a bad one. We call the creation of such a good value a *success* and speak of a *failure* otherwise. The *success probability* is then the probability of a success, equivalent to the pheromone on the respective edge. We first investigate how quickly the success probability increases if the algorithm only has successes for a while.

Definition 1. *Let p be the current success probability of a specific bit. Let $p^{(t)}$ be its success probability after $t \geq 0$ successes and no failures at the bit.*

The following lemma describes how the rewarded success probability relates to the unrewarded one.

Lemma 1. *For every $t \geq 0$, unless $p^{(t)}$ is capped by pheromone borders,*

$$p^{(t)} \ = \ 1-(1-p)\cdot(1-\rho)^t.$$

Proof. We prove the claim by induction on t. The case $t = 0$ is obvious. For $t \geq 1$, using the induction hypothesis,

$$\begin{aligned} p^{(t)} &= (1-\rho)p^{(t-1)} + \rho \\ &= (1-\rho)(1-(1-p)\cdot(1-\rho)^{t-1}) + \rho \\ &= 1-\rho-(1-p)\cdot(1-\rho)^{t} + \rho \\ &= 1-(1-p)\cdot(1-\rho)^{t}. \end{aligned}$$

□

By the preceding lemma, $p \geq q$ implies $p^{(t)} \geq q^{(t)}$ for all $t \geq 0$. Note that this also holds if the upper border for the success probability is hit. This justifies in our forthcoming analyses the places where actual success probabilities are replaced with lower bounds on these probabilities.

3.1 The Fitness-Level Method for the Analysis of ACO

In the following, we demonstrate how to derive upper bounds on the expected optimization time of MMAS* and MMAS on unimodal functions, especially OneMax and LeadingOnes. The techniques from this subsection were first presented by Gutjahr and Sebastiani (2008) in a more general setting, including arbitrary values for the pheromone borders. We stick to the presentation from Neumann et al (2009) where this method was adapted to MMAS* with our precise choice of pheromone bounds. Our presentation follows the presentation of the fitness-level method for evolutionary algorithms. This enables us to highlight the similarities between EAs and ACO and it reveals a way to directly transfer runtime bounds known for EAs to MMAS*. This is an important step towards a unified theory of EAs and ACO.

There is a similarity between MMAS* and evolutionary algorithms that can be exploited to obtain good upper bounds. Suppose that during a run there is a phase during which MMAS* never replaces the best-so-far solution x^*. This implies that the best-so-far solution is reinforced again and again until all pheromone values have reached their upper or lower borders corresponding to the setting of the bits in x^*. We can say that x^* has been "frozen in pheromone." The probability of creating a 1 for every bit is now either $1/n$ or $1-1/n$. The distribution of constructed solutions equals the distribution of offspring of the (1+1) EA* and (1+1) EA with x^* as the current search point. We conclude that, as soon as all pheromone values touch their upper or lower borders, MMAS* behaves like the (1+1) EA* until a solution with larger fitness is encountered. This similarity between ACO and EAs can be used to transfer the fitness-level method to ACO. In particular, upper bounds for MMAS* will be obtained from bounds for the (1+1) EA by adding the so-called freezing time described in the following.

Suppose the current best-so-far solution x^* is not changed and consider the random time t^* until all pheromones reach their borders corresponding to the bit values in x^*. We will refer to this random time as *freezing time*. Plugging the pheromone border $1 - 1/n$ into Lemma 1, along with the worst-case initial pheromone value $1/n$, and solving the resulting equation for t, we have that t^* is bounded from above by $-\ln(n-1)/\ln(1-\rho)$. We use $\ln(1-\rho) \le -\rho$ for $0 \le \rho \le 1$ and arrive at the handy upper bound

$$t^* \le \frac{\ln n}{\rho}. \tag{2}$$

We now use the freezing time t^* to derive a general upper bound on the expected optimization time of MMAS* by making use of the following restricted formulation of the fitness-level method. Let $f_1 < f_2 < \cdots < f_m$ be an enumeration of all fitness values and let A_i, $1 \le i \le m$, contain all search points with fitness f_i. In particular, A_m contains only optimal search points. Now, let s_i, $1 \le i \le m-1$, be a lower bound on the probability of the (1+1) EA (or, in this case equivalently, the (1+1) EA*) of creating an offspring in $A_{i+1} \cup \cdots \cup A_m$, provided the current population belongs to A_i. The expected waiting time until such an offspring is created is at most $1/s_i$ and then the set A_i is left for good. As every set A_i has to be left at most once, the expected optimization time for the (1+1) EA and the (1+1) EA* is bounded above by

$$\sum_{i=1}^{m-1} \frac{1}{s_i}. \tag{3}$$

Consider t^* steps of MMAS* and assume $x^* \in A_i$. Either the best-so-far fitness increases during this period or all pheromone values are frozen. In the latter case, the probability of creating a solution in $A_{i+1} \cup \cdots \cup A_m$ is at least s_i and the expected time until the best-so-far fitness increases is at most $t^* + 1/s_i$. We arrive at the following upper bound for MMAS*:

$$\sum_{i=1}^{m-1} \left(t^* + \frac{1}{s_i}\right).$$

This is a special case of Inequality (13) in Gutjahr and Sebastiani (2008). Using $t^* \le (\ln n)/\rho$, we obtain the more concrete bound

$$\frac{m \ln n}{\rho} + \sum_{i=1}^{m-1} \frac{1}{s_i}. \tag{4}$$

The right-hand sum is the upper bound obtained for the (1+1) EA and (1+1) EA* from (3). Applying the fitness-level method to MMAS*, we obtain upper bounds that are only by an additive term $(m \ln n)/\rho$ larger than the corresponding bounds for (1+1) EA and (1+1) EA*. This additional term results from the (pessimistic) assumption that on all fitness levels MMAS* cannot find a better solution until all pheromones are frozen. We will see examples where for large ρ this bound is of the same order of growth as the bound for (1+1) EA and (1+1) EA*. However, if ρ is very small, the bound for MMAS* typically grows large. This reflects the long time needed for MMAS* to move away from the initial random search and to focus on promising regions of the search space.

The following theorem has already been proven in Gutjahr and Sebastiani (2008) with a more general parametrization for the pheromone borders. We present a proof for MMAS* using our simplified presentation of the fitness-level method.

Theorem 1. *The expected optimization time of MMAS* on* ONEMAX *is bounded from above by* $O((n \log n)/\rho)$.

Proof. The proof is an application of the above-described fitness-level method with respect to the fitness-level sets $A_i = \{x \mid f(x) = i\}$, $0 \le i \le n$. On level A_i, a sufficient condition to increase the fitness is to flip a 0-bit and not to flip the other $n-1$ bits. For a specific 0-bit, this probability is $1/n \cdot (1-1/n)^{n-1} \ge 1/(en)$. As the events for all $n-i$ 0-bits are disjoint, $s_i \ge (n-i)/(en)$ and we obtain the bound

$$\sum_{i=0}^{n-1} \frac{en}{n-i} = en \sum_{i=1}^{n} \frac{1}{i} = O(n \log n).$$

Using (4), the upper bound $O((n \log n)/\rho)$ follows. □

The function LEADINGONES counts the number of leading ones in the considered bit string. A non-optimal solution may always be improved by appending a single one to the leading ones. We analyze MMAS* on LEADINGONES using our simplified presentation of the fitness-level method.

Theorem 2. *The expected optimization time of MMAS* on* LEADINGONES *is bounded from above by* $O(n^2 + (n \log n)/\rho)$.

Proof. For $0 \le i < n$ the (1+1) EA adds an $(i+1)$-st bit to the i leading ones of the current solution with probability $s_i = (1-1/n)^i \cdot 1/n \ge 1/(en)$. Using the bound (4) results in the upper bound $((n+1)\ln n)/\rho + en^2 = O(n^2 + (n \log n)/\rho)$. □

In the remainder of this section we review original results from Neumann et al (2009). First, we apply the fitness-level method to MMAS* on arbitrary unimodal functions. We also extend the method to yield upper bounds for MMAS on unimodal functions. This extension is not immediate as MMAS may switch between solutions of equal function value, which may prevent the pheromones from freezing. Moreover, we present lower bounds on the expected optimization time of MMAS* on all functions with unique optimum and an improved specialized lower bound for LEADINGONES. On one hand, these bounds allow us to conclude that the fitness-level method can provide almost tight upper bounds. On the other hand, as can be seen from a more detailed analysis in Section 3.4, the method still leaves room for improvement using specialized techniques.

3.2 Upper Bounds for Unimodal Functions

Unimodal functions are an important and well-studied class of fitness functions in the literature on evolutionary computation. A function is called *unimodal* if every non-optimal search point has a Hamming neighbor with strictly larger function value. Unimodal functions are often believed to be easy to optimize. This holds if the set of different function values is not too large. On the other hand, Droste, Jansen, and Wegener (2006) proved for classes of unimodal functions with many function values that *every* black-box algorithm needs exponential time on average.

We consider unimodal functions attaining d different fitness values for arbitrary $d \in \mathbb{N}$. Such a function is optimized by the (1+1) EA and (1+1) EA* in expected time $O(nd)$ (see Droste et al (2002)). This bound is transferred to MMAS* by the following theorem.

Theorem 3. *The expected optimization time of MMAS* on a unimodal function attaining d different function values is $O((n + (\log n)/\rho)d)$.*

Proof. Because of the unimodality there is for each current search point x a better Hamming neighbor x' of x in a higher fitness-level set. The probability for the (1+1) EA (or, equivalently, MMAS* with all pheromone values at a border) to produce x' in the next step is at least $1/(en)$. By (4), this completes the proof. □

In order to freeze pheromones after t^* steps without an improvement, it is essential that equally good solutions are rejected. The fitness-level argumentation, including the bound from (4), cannot directly be transferred to MMAS as switching between solutions of equal quality can prevent the system from freezing. Nevertheless, we are able to prove a similar upper bound on the optimization time of MMAS that is by a factor of n^2 worse than the bound for MMAS* in Theorem 3 if $\rho = O((\log n)/n)$. Despite the factor n^2, Theorem 4 yields a polynomial bound for MMAS if and only if Theorem 3 yields a polynomial bound for MMAS*.

Theorem 4. *The expected optimization time of MMAS on a unimodal function attaining d different fitness values is $O(((n^2 \log n)/\rho)d)$.*

Proof. We only need to show that the expected time for an improvement of the best-so-far solution is at most $O((n^2 \log n)/\rho)$. The probability that MMAS produces within $O((\log n)/\rho)$ steps a solution being at least as good as (not necessarily better than) the best-so-far solution x^* is $\Omega(1)$ since after at most $(\ln n)/\rho$ steps without exchanging x^* all pheromone values have touched their borders and then the probability of rediscovering x^* is $(1 - 1/n)^n = \Omega(1)$. We now show that the conditional probability of an improvement if x^* is replaced is $\Omega(1/n^2)$.

Let $x_1, \dots, x_m$ be an enumeration of all solutions with fitness values equal to the best-so-far fitness value. Because of the unimodality, each x_i, $1 \le i \le m$, has some better Hamming neighbor y_i; however, the y_i need not be disjoint. Let X and Y denote the event to generate some x_i or some y_i, respectively, in the next step. In the worst case $y_1, \ldots, y_m$ are the only possible improvements, hence the theorem follows if we can show $\text{Prob}(Y \mid X \cup Y) \geq 1/n^2$, which is implied by $\text{Prob}(Y) \geq \text{Prob}(X)/(n^2 - 1)$.

If $p(x_i)$ is the probability of constructing x_i then $p(x_i)/p(y_i) \leq (1 - \frac{1}{n})/\frac{1}{n} = n - 1$ as the constructions only differ in one bit. Each y_i may appear up to n times in the sequence $y_1, \ldots, y_m$, hence $\text{Prob}(Y) \geq \frac{1}{n} \sum_{i=1}^{m} p(y_i)$ and

$$\text{Prob}(X) = \sum_{i=1}^{m} p(x_i) \leq (n-1) \cdot \sum_{i=1}^{m} p(y_i) \leq n(n-1) \cdot \text{Prob}(Y).$$

Therefore, $\text{Prob}(Y) \geq \text{Prob}(X)/(n^2 - 1)$ follows. □

Theorems 3 and 4 show that the expected optimization times of both MMAS and MMAS* are polynomial for all unimodal functions as long as $d = \text{poly}(n)$ and $\rho = 1/\text{poly}(n)$.

3.3 A General Lower Bound

In order to judge the quality of upper bounds on expected optimization times it is helpful to aim at lower bounds. We present a lower bound that is weak but very general; it holds for both MMAS and MMAS* on all functions where the global optimum is unique.

Theorem 5. *Let $f\colon \{0,1\}^n \to \mathbb{R}$ be a function with a unique global optimum. Choosing $\rho = 1/\text{poly}(n)$, the expected optimization time of MMAS and MMAS* on f is $\Omega((\log n)/\rho - \log n)$.*

Proof. As 0-edges and 1-edges are treated symmetrically, we can assume w. l. o. g. that 1^n is the unique optimum. If, for each bit, the success probability (defined as the probability of creating a 1) is bounded from above by $1 - 1/\sqrt{n}$ then the solution 1^n is created with only an exponentially small probability of at most $(1 - 1/\sqrt{n})^n \leq e^{-\sqrt{n}}$. Using the uniform initialization and Lemma 1, the success probability of a bit after t steps is bounded from above by $1 - (1-\rho)^t/2$. Hence, all success probabilities are bounded as desired within $t := (1/\rho - 1) \cdot (\ln(n/4)/2) = \Omega((\log n)/\rho - \log n)$ steps since

$$1 - \frac{1}{2}(1-\rho)^t \leq 1 - \frac{e^{-(\ln n - \ln 4)/2}}{2} = 1 - \frac{1}{\sqrt{n}}.$$

Since $\rho = 1/\text{poly}(n)$ and, therefore $t = \text{poly}(n)$, the total probability of creating the optimum in t steps is still at most $te^{-\sqrt{n}} = e^{-\Omega(\sqrt{n})}$, implying the lower bound on the expected optimization time. □

There is often a trade-off between quality and generality of runtime bounds. The bounds presented so far are general, but they often do not represent the best possible bounds when considering specific functions. The following section shows exemplarily that stronger results can be obtained by more specialized investigations.

3.4 Specialized Analyses

The fitness-level method is a quite powerful and general method to derive upper bounds on the expected optimization time. However, as mentioned above, it relies on the pessimistic assumption that pheromones have to be frozen after each improvement of the function value. This provably leads to an overestimation of the expected runtime on the function LEADINGONES. Neumann et al (2009) proved the following bound which is much more precise.

Theorem 6. *The expected optimization time of MMAS and MMAS* on* LEADINGONES *is bounded by* $O(n^2 + n/\rho)$ *and* $O\left(n^2 \cdot (1/\rho)^{\varepsilon} + \frac{n/\rho}{\log(1/\rho)}\right)$ *for every constant* $\varepsilon > 0$.

The first bound is better than the one from Theorem 2 by a factor of $\Theta(\log n)$ if $\rho = O(1/n)$. In addition, the second bound basically saves another factor of $\Theta(\log n)$ if $\rho \leq 1/n^{1+\Omega(1)}$ holds, which, e. g., is the case if $\rho \leq 1/n^2$.

The proof of the improved bounds makes use of the following observation: As soon as a bit contributes to the LEADINGONES-value, i. e., the bit is part of the block of leading ones, its pheromone value will only increase in the following. (To decrease the pheromone, a solution with less leading ones would have to be accepted, which is ruled out by the selection.) Hence, if bits enter the block of leading ones one after the other with a certain delay, earlier gained bits will have had time to reach the maximal pheromone value. The freshly gained bits are likely to have some intermediate pheromone value, but if the delay between improvements is large enough, there is still a good probability of rediscovering all bits of the block of leading ones with a good probability. Basically, the analysis for the first bound from Theorem 6 shows that an average delay of $\Theta(1/\rho)$ steps between subsequent improvements is enough. Then there is always only a small "window" consisting of $O(\log n)$ bits in the block of leading ones where the pheromone values have not yet reached the upper borders. The bits in the window can be ranked from the left to the right, which means that the bits to the left have received more increases of their pheromone. Taking also this into consideration, a technical calculation proves the first bound from Theorem 6. Repeating these arguments for a smaller average delay of $\Theta(1/(\varepsilon\rho\ln(1/\rho))$ yields the second bound.

It is interesting that an almost tight lower bound can be derived. The following theorem shows that the expected optimization time of MMAS* on LEADINGONES is $\Omega(n^2 + n/(\rho\log(2/\rho)))$, hence never better than $\Omega(n^2)$. (We write $\log(2/\rho)$ in the lower bound instead of $\log(1/\rho)$ to make the bound $\Omega(n^2)$ for any constant ρ and to avoid division by 0.) Apart from this technical detail, the lower bound is tight with the upper bounds from Theorem 6 for $\rho = \Omega(1/n)$ and $\rho \leq n^{-(1+\Omega(1))}$, hence for almost all ρ.

Theorem 7. *Choosing $\rho = 1/\text{poly}(n)$, the expected optimization time of MMAS* on* LEADINGONES *is bounded from below by $\Omega\left(n^2 + \frac{n/\rho}{\log(2/\rho)}\right)$.*

The proof of this lower bound is the most demanding analysis in the paper by Neumann et al (2009). It makes use of tailored probabilistic tools for the analysis of a so-called martingale process. Such processes arise at the bits that have never contributed to the LEADINGONES-value so far. For this reason, their pheromone values are completely random, and, unless the bits become part of the block of leading ones, their expected values remain at their initial value of 1/2. This is exactly the property of a martingale. However, there are random fluctuations (a variance of the random change of pheromone) at these bits which drive their pheromone values away from the "middle" 1/2 by means of an unbiased random walk. During the time needed to gather the first $n/2$ leading ones, the pheromones of the rightmost $n/2$ bits therefore tend to reach one of their borders $1/n$ and $1 - 1/n$. Due to symmetry, on average half of these bits touch their lower border and have a good chance to remain there until the respective bit becomes the first 0-bit. Then an event of probability $1/n$ is necessary to "flip" the bit. As this situation is likely to occur a linear number of times, the bound $\Omega(n^2)$ follows. Finally, the bound $\frac{n/\rho}{\log(2/\rho)}$ goes back to an analysis on the average time between two improvements. Similarly as above, the "window" of bits with an intermediate pheromone value among the leading ones is studied. If the average time between two improvements is too small, the windows becomes too large and the probability of rediscovering the leading ones is too low to sustain the average time between two improvements.

Without going into the proof details, the summaries from this subsection should emphasize that specialized techniques for the analysis of ACO algorithms can lead to much improved results. However, much more involved proofs are required and the methods are not easy to transfer to different scenarios.

4 How ACO Algorithms Deal with Plateaus

Rigorous runtime analyses can be used to estimate the expected optimization time and to make precise predictions about the practical performance of ACO algorithms. This can be used to clarify important design issues from a rigorous perspective. One such issue already discussed in Gutjahr (2007); Gutjahr and Sebastiani (2008) is the question whether the current best-so-far solution should be replaced by a new solution with the same function value. This section again reviews results from Neumann et al (2009).

The general upper bounds from Theorems 3 and 4 for unimodal functions yield a gap of only polynomial size between MMAS and MMAS*. In addition, we have proven the same upper bounds on LEADINGONES for both MMAS and MMAS*. This may give the impression that MMAS and MMAS* behave similarly on all functions. However, this only holds for functions with a certain gradient towards better solutions. On plateaus MMAS and MMAS* can have a totally different behavior.

We consider the function NEEDLE where only one single solution has objective value 1 and the remaining ones get value 0. In its general form, the function is defined as

$$\text{NEEDLE}(x) := \begin{cases} 1 & \text{if } x = x_{\text{OPT}}, \\ 0 & \text{otherwise,} \end{cases}$$

where x_{OPT} is the unique global optimum. Gutjahr and Sebastiani (2008) compare MMAS* and (1+1) EA* with respect to their runtime behavior. For suitable values of ρ that are exponentially small in n, MMAS* has expected optimization time $O(c^n)$, $c \geq 2$ an appropriate constant, and beats the (1+1) EA*. The reason is that MMAS* behaves nearly as random search on the search space while the initial solution of the (1+1) EA* has Hamming distance n to the optimal one with probability 2^{-n}. To obtain from such a solution an optimal one, all n bits have to flip, which has expected waiting time n^n, leading in summary to an expected optimization time $\Omega((n/2)^n)$. In the following, we show a similar result for MMAS* if ρ decreases only polynomially with the problem dimension n.

Theorem 8. *Choosing* $\rho = 1/\text{poly}(n)$, *the optimization time of MMAS* on* NEEDLE *is at least* $(n/6)^n$ *with probability* $1 - e^{-\Omega(n)}$.

Proof. Let x be the first solution constructed by MMAS* and denote by x_{OPT} the optimal one. As it is chosen uniformly at random from the search space, the expected number of positions where x and x_{OPT} differ is $n/2$ and there are at least $n/3$ such positions with probability $1 - e^{-\Omega(n)}$ using Chernoff bounds. At these positions of x the "wrong" edges of the construction graph are reinforced as long as the optimal solution has not been obtained. This implies that the probability of obtaining the optimal solution in the next step is at most $2^{-n/3}$. After at most $t^* \leq (\ln n)/\rho$ (see Inequality (2)) iterations, the pheromone values of x have touched their borders provided x_{OPT} has not been obtained. The probability of having obtained x_{OPT} within a phase of t^* steps is at most $t^* \cdot 2^{-n/3} = e^{-\Omega(n)}$. Hence, the probability of producing a solution that touches its pheromone borders and differs from x_{OPT} in at least $n/3$ positions before producing x_{OPT} is $1 - e^{-\Omega(n)}$. In this case, the expected number of steps to produce x_{OPT} is $(n/3)^n$ and the probability of having reached this goal within $(n/6)^n$ steps is at most 2^{-n}. □

The probability of choosing an initial solution x that differs from x_{OPT} by n positions is 2^{-n}, and in this case, after all n bits have reached their corresponding pheromone borders, the probability of creating x_{OPT} equals n^{-n}. Using the ideas of Theorem 8, the following corollary can be proved which asymptotically matches the lower bound for the (1+1) EA* given in Gutjahr and Sebastiani (2008).

Corollary 1. *Choosing* $\rho = 1/\text{poly}(n)$, *the expected optimization time of MMAS* on* NEEDLE *is* $\Omega((n/2)^n)$.

It is well known that the (1+1) EA that accepts each new solution has expected optimization time $\Theta(2^n)$ on NEEDLE (see Garnier, Kallel, and Schoenauer, 1999;

Wegener and Witt, 2005) even though it samples with high probability in the Hamming neighborhood of the latest solution. On the other hand, MMAS* will have a much larger optimization time unless ρ is superpolynomially small (Theorem 8). Our aim is to prove that MMAS is more efficient than MMAS* and almost competitive to the (1+1) EA. The following theorem even shows that the expected optimization time of MMAS on NEEDLE is at most by a polynomial factor larger than the one of the (1+1) EA unless ρ is superpolynomially small. We sketch the proof ideas and refer the reader to Neumann et al (2009) for the full proof.

Theorem 9. *The expected optimization time of MMAS on* NEEDLE *is bounded from above by* $O((n^2 \log n)/\rho^2 \cdot 2^n)$.

Sketch of proof. By the symmetry of the construction procedure and uniform initialization, we w. l. o. g. assume that the needle x_{OPT} equals the all-ones string 1^n. As in Wegener and Witt (2005), we study the process on the constant function $f(x) = 0$. The first hitting times for the needle are the same on NEEDLE and the constant function, while the constant function is easier to study as MMAS accepts each new search point forever for this function.

The proof idea is to study a kind of "mixing time" $t(n)$ after which each bit is independently set to 1 with a probability of at least 1/2 regardless of its initial success probability (recall that this means the probability of setting the bit to 1). Since bits are treated independently, this implies that the probability of creating the needle is at least 2^{-n} in some step after at most $t(n)$ iterations. We successively consider independent phases of (random) length $t(n)$ until the needle is sampled. The number of phases required follows a geometric distribution with parameter at least 2^{-n}, hence, the expected number of phases required to sample the needle is at most 2^n. By the linearity of expectation, the expected time until 2^n phases have elapsed is bounded by $\mathrm{E}(t(n)) \cdot 2^n$. The theorem follows if we can show that $\mathrm{E}(t(n)) = O((n^2 \log n)/\rho^2)$.

In order to bound $t(n)$, we study a random walk on the success probabilities. Consider the independent success probabilities of the n bits for any initial distribution. We call a success probability *good* at a certain time t if is has been bounded from below by 1/2 at least once in the t steps after initialization and *bad* otherwise. We are interested in the time T^* until all n success probabilities have become good. For each single success probability, the expected time until becoming good is $O(n^2/\rho^2)$ (see Neumann et al (2009)). Due to Markov's inequality, the time is $O(n^2/\rho^2)$ with probability at least 1/2. Repeating $2 \log n$ independent such phases, this implies that after $O(n^2/\rho^2 \cdot \log n)$ steps, each success probability is bad with probability at most $1/(2n)$. Hence, by the union bound, the probability is only at most 1/2 that there is a bad success probability left after this number of steps. Repeating this argument an expected number of at most 2 times, $\mathrm{E}(T^*) = O((n^2 \log n)/\rho^2)$ follows. By definition, all success probabilities have been at least 1/2 at least once after T^* steps. One can show that then the probability of creating bit value 1 for such a bit remains at least 1/2. Hence, T^* is an upper bound on $t(n)$, and the theorem follows. □

The function NEEDLE requires an exponential optimization time for each algorithm that has been considered. Often plateaus are much smaller, and randomized search heuristics have a good chance to leave them within a polynomial number of steps. Gutjahr and Sebastiani (2008) consider the function NH-ONEMAX that consists of the NEEDLE-function on $k = \log n$ bits and the function ONEMAX on $n - k$ bits, which can only be optimized if the needle has been found on the needle part. The function is defined as

$$\text{NH-ONEMAX}(x) = \left(\prod_{i=1}^{k} x_i\right)\left(\sum_{i=k+1}^{n} x_i\right).$$

Taking into account the logarithmic size of the NEEDLE-function of NH-ONEMAX, MMAS* with polylogarithmically small ρ cannot optimize the needle part within an expected polynomial number of steps. The proof ideas are similar to those used in the proof of Theorem 8. After initialization the expected Hamming distance of the needle part to the needle is $(\log n)/2$, and it is at least $(\log n)/3$ with probability $1-o(1)$. Working under this condition, this means that the probability of sampling the needle is at most $2^{-(\log n)/3} = n^{-1/3}$ in each step before the needle is found. As $\rho = 1/\text{polylog}(n)$ holds, the lower pheromone borders of the $(\log n)/3$ "wrong" bits from the initial solution are reached in at most $t^* \leq (\ln n)/\rho = \text{polylog}(n)$ steps. Hence, the needle is found before this situation has been reached with probability at most $\text{polylog}(n)/n^{1/3} = o(1)$. Afterwards, the probability of sampling the needle is at most $n^{-(\log n)/3} = 2^{-\Omega(\log^2 n)}$. This proves the following superpolynomial lower bound.

Theorem 10. *If $\rho = 1/\text{polylog}(n)$, the expected optimization time of MMAS* on* NH-ONEMAX *is* $2^{\Omega(\log^2 n)}$.

Also the proof of Theorem 9 carries over to a major extent. The random walk arguments leading to $\text{E}(t(n)) = O((n^2 \log n)/\rho^2)$ still hold since random bits are considered independently and the borders for the pheromone values have not been changed. What has been changed is the size of the needle. As now the needle part only consists of $\log n$ bits, the probability of creating it is at least $2^{-\log n} = 1/n$ after $t(n)$ steps. Hence, MMAS can find the needle after an expected number of $O((n^2 \log n)/\rho^2 \cdot n)$ steps. After this goal has been achieved, the unimodal function ONEMAX, which contains at most $n+1$ different function values, has to be optimized. We conclude from Theorem 4, the general bound on unimodal functions, that MMAS optimizes ONEMAX in an expected number of $O((n^3 \log n)/\rho)$ steps. Putting the two bounds together, the following result has been proved.

Theorem 11. *The expected optimization time of MMAS on* NH-ONEMAX *is at most $O((n^3 \log n)/\rho^2)$.*

This bound is polynomial if $\rho = 1/\text{poly}(n)$, in contrast to the superpolynomial bound for MMAS* from Theorem 10. Hence, MMAS is superior to MMAS* on NH-ONEMAX as well.

5 The Effect of Hybridizing ACO with Local Search

In this section we now turn to the hybridization of ACO with local search and follow investigations carried out by Neumann, Sudholt, and Witt (2008). Combining ACO with local search methods is quite common (see, e. g., Dorigo and Stützle, 2004; Hoos and Stützle, 2004; Levine and Ducatelle, 2004). Experimental investigations show that the combination of ACO with a local search procedure improves the performance significantly. On the other hand, there are examples where local search cannot help to improve the search process or even mislead the search process (Balaprakash, Birattari, Stützle, and Dorigo, 2006). Therefore, it is interesting to figure out how the incorporation of local search into ACO algorithms can significantly influence the optimization process.

Our aim is to point out situations where the effect of local search becomes visible in a way that can be tackled by rigorous arguments. Therefore we present functions where MMAS variants with and without local search show a strongly different runtime behavior. On one function, MMAS with local search outperforms MMAS without local search, while on a different function the effect is reversed. The differences are so drastic that the question of whether to use local search or not decides between polynomial and exponential runtimes.

We enhance MMAS* with local search and call the result MMAS-LS*. In the following, LocalSearch(x) is a procedure that, starting from x, repeatedly replaces the current solution by a Hamming neighbor with a strictly larger function value until a local optimum is found. We do not specify a pivot rule, hence we implicitly deal with a class of algorithms.

Algorithm 2. MMAS-LS*

Set $\tau(e) := 1/2$ for all $e \in E$.
Construct a solution x^*.
Set $x^* :=$ LocalSearch(x^*).
Update pheromones with respect to x^*.
repeat
    Construct a solution x.
    Set $z :=$ LocalSearch(x).
    if $f(z) > f(x^*)$ **then** $x^* := z$.
    Update pheromones with respect to x^*.

When taking the number of function evaluations as performance measure, the computational effort of local search has to be accounted for. The objective functions considered in the following only have a linear number of function values, hence the number of function evaluations in one local search call is bounded by $O(n)$. Depending on the pivot rule, the number of evaluations needed to find a better Hamming neighbor may vary; however, it is trivially bounded by n. Hence, the number of function evaluations is at most by a factor $O(n^2)$ larger than the number of iterations.

Hybridization with local search has become very popular for various other optimization paradigms such as estimation-of-distribution algorithms (Aickelin, Burke, and Li, 2007) and evolutionary algorithms. Evolutionary algorithms using local search are known as *memetic (evolutionary) algorithms.* They have been successfully applied to many combinatorial problems, see the book by Hart, Krasnogor, and Smith (2004) and the survey by Krasnogor and Smith (2005). One particular memetic algorithm is known *iterated local search* (Lourenço, Martin, and Stützle, 2002) where local search is used in every generation to turn new search points into local optima. The first rigorous runtime analyses for memetic algorithms were presented by Sudholt (2009). The author highlights the importance of a good balance between local search and evolutionary search. He presents examples where small changes to the parametrization have a huge impact on the performance of simple memetic algorithms. A further study for problems from combinatorial optimization (Sudholt, 2008) demonstrates that an iterated local search algorithm with a sophisticated local search can outperform several common heuristics on simply structured problem instances.

In the rest of this chapter, we want to examine the effect of combining ACO algorithms with local search methods. Neumann et al (2008) have pointed out that the effect of using local search with ACO algorithms is manifold. Firstly, local search can help to find good solutions more quickly as it increases the "greediness" within the algorithm. As argued in Sudholt (2009) for memetic algorithms, local search can also be used to discover the real "potential" of a solution as it can turn a bad looking solution into a good local optimum. Moreover, the pivot rule used in local search may guide the algorithm towards certain regions of the search space. For example, first ascent pays more attention to the first bits in the bit string, which may induce a search bias. However, we will not deal with this effect here. In particular, our functions are designed such that the pivot rule is not essential.

There is another effect that has been investigated more closely in Neumann et al (2008). The pheromone values induce a sampling distribution over the search space. On typical problems, once the best-so-far solution has reached a certain quality, sampling new solutions with a high variance becomes inefficient and the current best-so-far solution x^* is maintained for some time. The analyses from Section 3 have shown that then the pheromones quickly reach the upper and lower borders corresponding to x^*. This means that the algorithm turns to sampling close to x^*. In other words, MMAS variants typically reach a situation where the "center of gravity" of the sampling distribution follows the current best-so-far solution and the variance of the sampling distribution is low.

When introducing local search into an MMAS algorithm, this may not be true. Local search is able to find local optima that are far away from the current best-so-far solution. In this case the "center of gravity" of the sampling distribution is far away from the best-so-far solution.

Assume there is a path of Hamming neighbors with increasing function value leading to a local optimum. Assume further that all points close to the path have

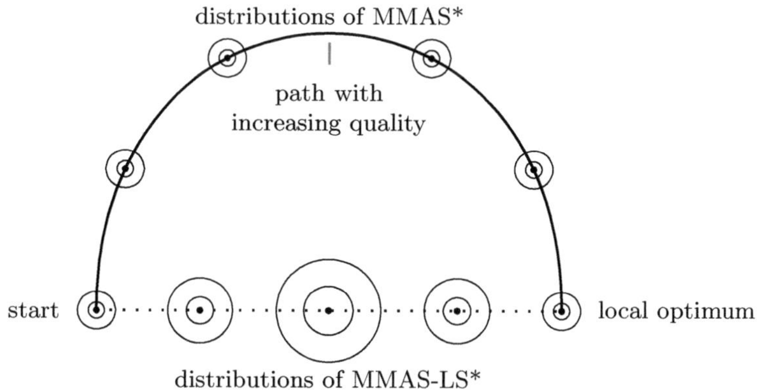

Fig. 3. A sketch of the search space showing the behavior of MMAS* and MMAS-LS*. The dots and circles indicate the sampling distributions of MMAS* and MMAS-LS*, respectively, at different points of time. While the distribution of MMAS* tends to follow the path of increasing function value from left to right, the distribution of MMAS-LS* takes a direct route towards the local optimum.

lower quality. Then for MMAS* it is likely that the sampling distribution closely follows the path. The path of increasing function value need not be straight. In fact, it can make large bends through the search space until a local optimum is reached. On the other hand, MMAS-LS*, when starting with the same setting, will reach the local optimum within a single iteration of local search. Then the local optimum becomes the new best-so-far solution x^*, while the sampling distribution is still concentrated around the starting point. In the following iterations, as long as the best-so-far solution is not exchanged, the pheromone values on all bits synchronously move towards their respective borders in x^*. This implies for the sampling distribution that the "center of gravity" takes a (sort of) direct route towards the local optimum, irrespective of the bent path taken by local search. An illustration is given in Figure 3.

Consequences are that different parts of the search space are sampled by MMAS* and MMAS-LS*, respectively. Moreover, with MMAS* the variance in the solution construction is always quite low as the sampling distribution is concentrated on certain points on the path. But when the best-so-far solution with local search suddenly moves a long distance, the variance in the solution construction may be very high as the bits differing between the starting point and x^* may have pheromones close to $1/2$. These bits are assigned almost randomly. This strongly resembles a uniform crossover operation well known in evolutionary computation. There, every bit in the offspring receives a bit value from a parent that is chosen uniformly at random and anew for each bit, which implies that bits differing in the two parents are assigned randomly. MMAS-LS* in this setting therefore simulates a uniform crossover of the starting point and the local optimum x^*.

Our aim in the following is to create functions where MMAS* and MMAS-LS* have a different runtime behavior. Moreover, we want the performance difference to be drastic in order to show how deep the impact of local search can possibly be. To this end, we exploit that the sampling distributions can follow different routes through the search space. For one function we place a target region with many global optima on the straight line between starting point and local optimum and turn the local optimum into a trap that is hard to overcome. In such a setting, we expect MMAS-LS* to drastically outperform MMAS*. Contrarily, if the region of global optima is made a trap region and a new global optimum is placed close to the former trap, we expect MMAS-LS* to get trapped and MMAS* to find the global optimum effectively.

5.1 A Function Where Local Search Is Beneficial

We now formally define a function where local search is beneficial according to the ideas described above. It is named SP-Target (short path with target). The path with increasing function value is given by the set $\mathrm{SP} = \{1^i 0^{n-i} \mid 0 \leq i \leq n\}$. The path ends with the local optimum 1^n. Let $|x|_1$ denote the number of 1-bits in x and $|x|_0$ denote the number of 0-bits. A large target area containing all global optima is specified by $\mathrm{OPT} = \{x \mid |x|_1 \geq (3/4) \cdot n \wedge H(x, \mathrm{SP}) \geq n/(\gamma \log n)\}$, where $H(x, \mathrm{SP})$ denotes the Hamming distance of x to the closest search point of SP and $\gamma \geq 1$ is a constant to be chosen later. For all remaining search points, the function SP-Target gives hints to reach 0^n, the start of the path SP.

$$\text{SP-Target}(x) := \begin{cases} |x|_0 & \text{if } x \notin (\mathrm{SP} \cup \mathrm{OPT}), \\ n+i & \text{if } x = 1^i 0^{n-i} \in \mathrm{SP}, \\ 3n & \text{if } x \in \mathrm{OPT}. \end{cases}$$

The function SP-Target is sketched in Figure 4. Note that we have actually defined a class of functions dependent on γ. All following results will hold for arbitrary constant $\gamma \geq 1$ unless stated otherwise.

The following theorem shows that MMAS* without local search is not successful. We restrict ourselves to polynomially large $1/\rho$ here and also in the following as otherwise the ACO component would be too close to random search.

Theorem 12. *Choosing $\rho = 1/\mathrm{poly}(n)$, the optimization time of MMAS* on* SP-Target *is at least $2^{cn^{2/9}}$ with probability $1 - 2^{\Omega(n^{2/9})}$ for some constant $c > 0$.*

To prove the preceding theorem, we have to take into account situations where the pheromone values of MMAS* have not yet reached their borders and the construction procedure samples with high variance. This is the case in particular after initialization. The following lemma will be used to check the probability of finding the optimum in the early steps of MMAS* on SP-Target.

Lemma 2. *If the best-so-far solution of MMAS* has never had more than $2n/3$ 1-bits, the probability of creating a solution with at least $3n/4$ 1-bits is $2^{-\Omega(n)}$ in each iteration.*

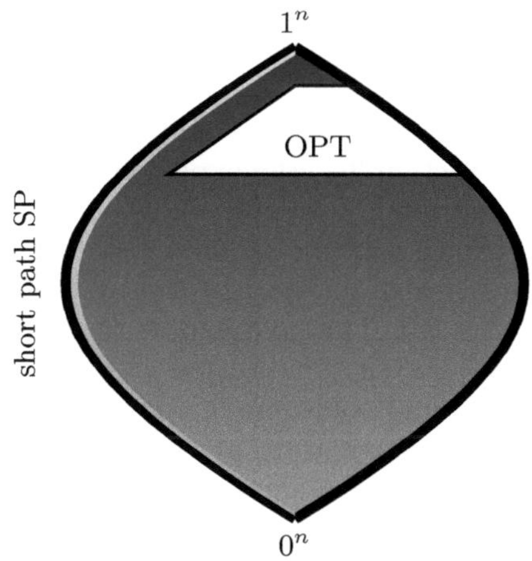

Fig. 4. Illustration of the Boolean hypercube and the function SP-Target. The vertical position of a search point is determined by the number of 1-bits. Its horizontal position is determined by the position of 1-bits in the bit string. The objective value is indicated by the brightness; dark areas indicate low values and light areas indicate high values.

Proof. The proof is an application of Chernoff bounds with respect to the number of ones in the solutions created by MMAS*. Let the potential $P_t := p_1 + \cdots + p_n$ at time t denote the current sum of the probabilities of sampling ones over all bits, which, by definition of the construction procedure, equals the expected number of ones in the next constructed solution. Observe that $P_t \leq 2n/3$ implies by Chernoff bounds that the probability of creating a solution with at least $3n/4$ 1-bits is $2^{-\Omega(n)}$. We now show: if all best-so-far solutions up to time t have at most $2n/3$ ones, then $P_i \leq 2n/3$ for $0 \leq i \leq t$. This will prove the lemma.

For the last claim, we denote by k the number of ones in the best-so-far solution according to which pheromones are updated. Due to the pheromone update mechanism, the new potential P_{i+1} is obtained from P_i and k according to $P_{i+1} = (1-\rho)P_i + k\rho$. Hence, if $P_i \leq 2n/3$ and $k \leq 2n/3$ then also $P_{i+1} \leq 2n/3$. The claim follows by induction since $P_0 = n/2 \leq 2n/3$. □

Proof of Theorem 12. We distinguish two phases in the run according to the best-so-far solution x^*. Phase 1 holds as long as $x^* \notin \text{SP}$ and $|x^*|_1 \leq 2n/3$, and Phase 2 applies as long as $x^* \in \text{SP}$. Our aim is to show that a typical run passes through the two phases in their order with a failure probability of $2^{-\Omega(n^{2/9})}$. The probability of finishing the second phase will be bounded by $2^{-\Omega(n^{2/9})}$ for each step of the phase. This implies the theorem as, by the union bound, the total probability in $2^{cn^{2/9}}$ iterations, $c > 0$ a small constant, is still $2^{-\Omega(n^{2/9})}$.

Consider the first (and best-so-far) solution x^* created by MMAS*. By Chernoff bounds, $n/3 \leq |x^*|_1 \leq 2n/3$ with probability $1 - 2^{-\Omega(n)}$. There is only a single solution in SP for each value of $|x^*|_1$. By the symmetry of the construction procedure, we conclude $\text{Prob}(x^* \in \text{SP} \mid |x^*|_1 = k) = 1/\binom{n}{k}$. The last expression is $2^{-\Omega(n)}$ for $n/3 \leq k \leq 2n/3$. Hence, with probability $1-2^{-\Omega(n)}$, there is a non-empty Phase 1. By Lemma 2, the probability that a specific iteration in Phase 1 creates an optimum is $2^{-\Omega(n)}$. Otherwise, the behavior is as for MMAS* on the function $|x|_0$. Using $\rho = 1/\text{poly}(n)$ and the analyses for the symmetric function $\text{OneMax}(x) = |x|_1$ from Section 3.1, the expected time until the first phase is finished is polynomial. The total failure probability in Phase 1 is bounded by the product of its expected length and the failure probability in a single iteration. Therefore, the total failure probability for the first phase is still of order $2^{-\Omega(n)}$.

In Phase 2 we have $x^* \in \text{SP}$. The goal is now to show that a solution from SP with high probability can only be created if the sampling distribution is sufficiently concentrated around solutions in SP. This in turn makes creating solutions of high Hamming distance from SP, including OPT, very unlikely.

We make this idea precise and consider a point $1^i0^{n-i} \in \text{SP}$. This search point consists of a prefix of i ones and a suffix of $n-i$ zeros. For a newly constructed solution x we define $P(i) := p_1 + \cdots + p_i$ as the expected number of ones in the prefix and $S(i) := (1-p_{i+1}) + \cdots + (1-p_n)$ as the expected number of zeros in the suffix. The number of ones in the prefix plus the number of zeros in the suffix yields the number of bits equaling in 1^i0^{n-i} and x, i. e., $n - H(1^i0^{n-i}, x)$. We call $P(i)$ ($S(i)$) *insufficient* if and only if $P(i) \leq i - i^{2/3}$ ($S(i) \leq (n-i) - (n-i)^{2/3}$) holds. We now show that with insufficiencies it is very unlikely to create 1^i0^{n-i}. As this holds for all i, we conclude that if SP is reached after a certain number of iterations, the pheromones do not have insufficiencies, with high probability.

Let $s(i)$ denote the probability of constructing the solution 1^i0^{n-i}. We distinguish three cases and apply Chernoff bounds to prove the following implications:

Case 1: $i < n^{2/3}$. Then insufficient $S(i)$ implies $s(i) = 2^{-\Omega(n^{1/3})}$.

Case 2: $i > n - n^{2/3}$. Then insufficient $P(i)$ implies $s(i) = 2^{-\Omega(n^{1/3})}$.

Case 3: $n^{2/3} \leq i \leq n - n^{2/3}$. Then insufficient $P(i)$ and insufficient $S(i)$ each imply $s(i) = 2^{-\Omega(n^{2/9})}$.

We assume that the described insufficiencies do not occur whenever a best-so-far solution $x^* = 1^i0^{n-i}$ in Phase 2 is accepted. The failure probability is $2^{-\Omega(n^{2/9})}$ for each new best-so-far solution x^*. Iterations in between two exchanges of x^* cannot create insufficiencies as $P(i)$ and $S(i)$ can only increase as long as x^* is maintained. Hence, we do not have insufficiencies in Phase 2 for at least $2^{\Omega(n^{2/9})}$ iterations with probability at least $1 - 2^{-\Omega(n^{2/9})}$.

Being in Phase 2 without insufficiencies, we show depending on the three cases for the current $x^* = 1^i0^{n-i}$ that creating an optimal solution has probability $2^{-\Omega(n^{2/9})}$. In the first case, the expected number of zeros in the suffix of x is at least $(n-i) - (n-i)^{2/3}$. By Chernoff bounds, the random number of zeros is at least $(n-i) - 2(n-i)^{2/3}$ with probability at least $1 - 2^{-\Omega(n^{1/3})}$. Along with $i < n^{2/3}$, it follows that then the solution has Hamming distance at most $3n^{2/3}$ from SP. By the definition of SP-Target, this is not enough to reach OPT. The second case is treated analogously. In the third case, the probability of obtaining less than $i - 2i^{2/3}$ ones in the prefix or less than $(n-i) - 2(n-i)^{2/3}$ zeros in the suffix is altogether bounded by $2^{-\Omega(n^{2/9})}$. Then the solution has

Hamming distance at most $4n^{2/3}$ from SP, which is also not enough to reach the optimum. This finishes the analysis of the second phase, and, therefore, proves the theorem. □

The following theorem proves the benefits of local search. Recall that the number of function evaluations is at most by a factor of $O(n^2)$ larger than the stated optimization time.

Theorem 13. *Choosing* $1/\text{poly}(n) \leq \rho \leq 1/16$, *the optimization time of MMAS-LS* on* SP-Target *is* $O(1/\rho)$ *with probability* $1 - 2^{-\Omega(n)}$. *If* $\gamma \geq 1$ *is chosen large enough but constant, the expected optimization time is also* $O(1/\rho)$.

Proof. Note that every call of local search ends either with 1^n or with a global optimum. If the initial local search creates $x^* = 1^n$, all pheromone values increase simultaneously and uniformly from their initial value 1/2 towards their upper border $1-1/n$. We divide a run into two phases. The first phase ends when either all pheromones become larger than 27/32 or when a global optimum has been found. The second phase ends when a global optimum has been found, hence it is empty if the first phase ended with an optimum.

We bound the length of the first phase by the first point of time t^* where all pheromone values exceed 27/32. By Lemma 1 after t steps the pheromone values are at least $\min\{1 - 1/n,\ 1 - (1/2)(1 - \rho)^t\}$. Solving the equation

$$1 - \left(\frac{1}{2}\right)(1 - \rho)^t \;=\; 27/32 \iff (1 - \rho)^t \;=\; 5/16$$

yields the upper bound

$$t^* \;\leq\; \left\lceil \frac{\ln(5/16)}{\ln(1 - \rho)} \right\rceil \;\leq\; \frac{\ln(16/5)}{\rho} + 1 \;=\; O(1/\rho).$$

The assumption $\rho \leq 1/16$ implies that at the last step in the first phase the pheromone value at every bit is within the interval $[25/32, 27/32]$, pessimistically assuming that a global optimum has not been found before (neither by a constructed ant solution, nor by local search). The next constructed search point x then fulfills the following two properties with probability $1 - O(2^{-n/2400})$:

1. $\frac{3n}{4} \leq |x|_1 \leq \frac{7n}{8}$,
2. $H(x, \text{SP}) \geq n/(\gamma \log n)$.

Using Chernoff bounds with $\delta := 1/25$, the failure probability for the first event is at most $2e^{-(25n/32)(\delta^2/3)} = 2e^{-n/2400}$. To bound the failure probability of the second event, given the first event, we exploit that all pheromone values are equal. Therefore, if we know that $|x|_1 = k$ then x is uniform over all search points with k ones. Since the number of search points with k ones is monotone decreasing for $3n/4 \leq k \leq 7n/8$, we only consider search points with $k = 7n/8$ ones as a worst case. The number of such search points is $\binom{n}{n/8}$, and the number of search points of Hamming distance at most $m := n/(\gamma \log n)$ from SP is at most $m \cdot \binom{n}{m}$. Altogether, the probability of $H(x, \mathrm{SP}) \le m$, given that $3n/4 \le |x|_1 \le 7n/8$, is bounded from above by

$$\frac{m\binom{n}{m}}{\binom{n}{n/8}} \le \frac{m\left(\frac{en}{m}\right)^m}{\left(\frac{n}{n/8}\right)^{n/8}} \le m \cdot 2^{o(n)} \cdot 8^{-n/8}.$$

The last expression is even $O(2^{-n/8})$. Altogether, the sum of the failure probabilities is $O(2^{-n/2400})$ as suggested, and the first statement follows.

For the second statement we estimate the time in the second phase, provided that the first phase has been unsuccessful. Using the bound $(\ln n)/\rho$ on the expected freezing time from Section 3.1 and $\rho = 1/\mathrm{poly}(n)$, the time to reach the pheromone border is $O((\log n)/\rho) = \mathrm{poly}(n)$, or an optimum is created anyway. With all pheromones at the upper border, the solution construction process equals a standard mutation of 1^n, i. e., flipping each bit in 1^n independently with probability $1/n$. Flipping the first m bits results in a global optimum as $0^m 1^{n-m}$ has Hamming distance at least m to $1^i 0^{n-i}$ for every i. The probability of creating $0^m 1^{n-m}$ in a standard mutation is at least

$$\left(\frac{1}{n}\right)^{n/(\gamma \log n)} \left(1 - \frac{1}{n}\right)^{n-n/(\gamma \log n)} \ge e^{-1} \cdot 2^{-n/\gamma}.$$

This means that the expected time in the second phase is $O(\mathrm{poly}(n) 2^{n/\gamma})$. Using that the first phase is unsuccessful only with probability $O(2^{-n/2400})$ and applying the law of total probability, the expected optimization time altogether is $O(1/\rho) + O(2^{-n/2400}) \cdot O(\mathrm{poly}(n) 2^{n/\gamma})$. The latter is $O(1)$ for $\gamma > 2400$, which proves the second statement. □

5.2 A Function Where Local Search Is Detrimental

Similarly to the function SP-Target, we design another function SP-Trap (short path with trap) where local search is detrimental, using ideas from Section 5. We take over the path with increasing function value, $\mathrm{SP} = \{1^i 0^{n-i} \mid 0 \le i \le n\}$, but in contrast to SP-Target, the former region of global optima now becomes a trap, $\mathrm{TRAP} = \{x \mid |x|_1 \ge (3/4) \cdot n \wedge H(x, \mathrm{SP}) \ge n/\log n\}$. The unique global optimum is placed within distance 2 from the local optimum: $\mathrm{OPT} = \{0^2 1^{n-2}\}$. This ensures that local search climbing the path SP cannot reach the global optimum. All remaining search points give hints to reach the start of the path.

$$\text{SP-Trap}(x) := \begin{cases} |x|_0 & \text{if } x \notin (\mathrm{SP} \cup \mathrm{TRAP} \cup \mathrm{OPT}), \\ n + i & \text{if } x = 1^i 0^{n-i} \in \mathrm{SP}, \\ 3n & \text{if } x \in \mathrm{TRAP}, \\ 4n & \text{if } x \in \mathrm{OPT}. \end{cases}$$

The function SP-Trap is sketched in Figure 5.

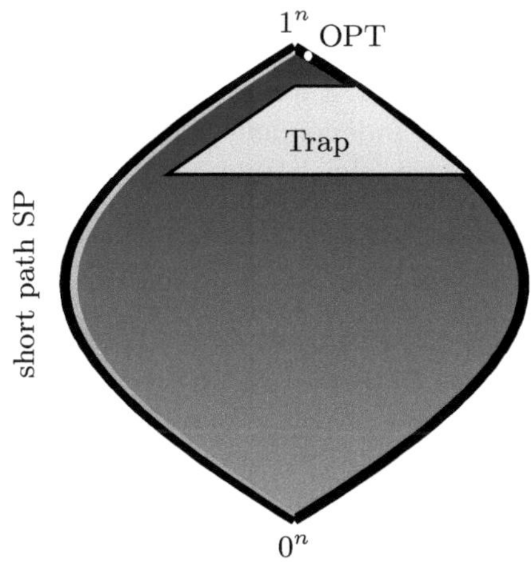

Fig. 5. Illustration of the Boolean hypercube and the function SP-Trap. The vertical position of a search point is determined by the number of 1-bits. Its horizontal position is determined by the position of 1-bits in the bit string. The objective value is indicated by the brightness; dark areas indicate low values and light areas indicate high values.

In the remainder of this section, we prove that MMAS* is efficient on SP-Trap while MMAS-LS* fails dramatically. Tuning the definition of SP-Trap, we could also extend the following theorem by a polynomial bound on the expected optimization time. We refrain from such modifications to illustrate the main effects.

Theorem 14. *Choosing $\rho = 1/poly(n)$, the optimization time of MMAS* on* SP-Trap *is $O((n \log n)/\rho + n^3)$ with probability* $1 - 2^{-\Omega(n^{2/9})}$.

Proof. By the argumentation from the proof of Theorem 12, the probability that a solution in TRAP is produced within $O(n^3)$ iterations is at most $2^{-\Omega(n^{2/9})}$.

Under the assumption that TRAP is never reached until the global optimum is found, MMAS* behaves equally on SP-Trap and a modified function where $x \in$ TRAP receives value $|x|_0$. We apply the fitness-level method described in Section 3.1 to estimate the expected optimization time on the latter, easier function. The number of fitness levels is $O(n)$. On every fitness level, the number of iterations until either all pheromones are frozen or the current best-so-far solution has improved is bounded by $(\ln n)/\rho$ with probability 1. We pessimistically assume that an improvement can only happen once all pheromones have been frozen. Then the optimization time is bounded by $O((n \log n)/\rho)$ plus the sum of waiting times for improvements on all fitness levels. Showing that the latter quantity is bounded by $O(n^3)$ with probability $1 - 2^{-\Omega(n)}$ completes the proof.

After freezing, the probability for an improvement from $x^* = 1^n$ equals $(1/n)^2 \cdot (1-1/n)^{n-2} \geq 1/(en^2)$. For all other $x^* \notin$ TRAP, there is always a better Hamming neighbor, hence the probability for an improvement is at least $1/(en)$.

Together, the expected waiting times for improvements on all fitness levels sum up to $en^2 + O(n) \cdot en = O(n^2)$. By Markov's inequality the probability of waiting more than cn^2 steps is at most 1/2 for a suitable constant $c > 0$. Hence, the probability that more than n independent phases of length cn^2 are needed is bounded by $2^{-\Omega(n)}$. Therefore, the bound $O(n^3)$ holds with probability $1 - 2^{-\Omega(n)}$. □

Theorem 15. *Choosing* $1/\text{poly}(n) \leq \rho \leq 1/16$, *the optimization time of MMAS-LS* on* SP-Trap *is at least* 2^{cn} *with probability* $1 - 2^{-\Omega(n)}$ *for some constant* $c > 0$.

Proof. We follow the lines of the proof of Theorem 13. As long as OPT $= 0^2 1^{n-2}$ is not created, the behavior of MMAS-LS* on SP-Trap and SP-Target is identical. Reconsider the first phase described in the proof of Theorem 13 (with the former OPT replaced by TRAP) and denote by $P := p_1 + \cdots + p_n$ the sum of probabilities of sampling ones over all bits. Throughout the phase, $P \leq 27n/32$, hence the probability of sampling at least $n - 2$ ones, which is necessary to reach OPT, is $2^{-\Omega(n)}$ according to Chernoff bounds.

With probability $1 - 2^{-\Omega(n)}$, the first best-so-far solution 1^n is replaced by some $x^{**} \in$ TRAP where $|x^{**}|_1 \leq 7n/8$ when the first phase is ended. Due to strict selection, x^{**} then can only be replaced if OPT is created. The latter has probability $2^{-\Omega(n)}$ for the following reasons: the P-value is at most $27n/32 \leq 7n/8$ when x^{**} is accepted. Hence, following the argumentation from the proof of Lemma 2, the P-value will not exceed $7n/8$ unless x^{**} is replaced. With a P-value of at most $7n/8$, creating OPT has probability $2^{-\Omega(n)}$ and the claim follows for a suitable constant $c > 0$. □

6 Conclusions

Ant colony optimization is a powerful metaheuristic that has found many applications for combinatorial and adaptive problems. In contrast to the rich number of successful applications, the theoretical understanding lags far behind practical success. A solid theoretical foundation is necessary to get a rigorous understanding of how ACO algorithms work. We have shown how simple ACO algorithms can be analyzed with respect to their computational complexity on example functions with different properties. This enables practitioners to gain new insights into their dynamic behavior and to clarify design issues, so that better algorithms can be developed. In particular, we have addressed the question whether the best-so-far solution should be replaced by new solutions of the same quality. As it is common practice to hybridize ACO with local search, we have discussed possible effects of introducing local search from a theoretical perspective and pointed out situations where the use of local search is either provably beneficial or provably disastrous.

References

Aickelin, U., Burke, E.K., Li, J.: An estimation of distribution algorithm with intelligent local search for rule-based nurse rostering. Journal of the Operational Research Society 58, 1574–1585 (2007)

Attiratanasunthron, N., Fakcharoenphol, J.: A running time analysis of an ant colony optimization algorithm for shortest paths in directed acyclic graphs. Information Processing Letters 105(3), 88–92 (2008)

Balaprakash, P., Birattari, M., Stützle, T., Dorigo, M.: Incremental local search in ant colony optimization: Why it fails for the quadratic assignment problem. In: Dorigo, M., Gambardella, L.M., Birattari, M., Martinoli, A., Poli, R., Stützle, T. (eds.) ANTS 2006. LNCS, vol. 4150, pp. 156–166. Springer, Heidelberg (2006)

Cormen, T.H., Leiserson, C.E., Rivest, R.L., Stein, C.: Introduction to Algorithms, 2nd edn. MIT Press, Cambridge (2001)

Doerr, B., Johannsen, D.: Refined runtime analysis of a basic ant colony optimization algorithm. In: Proceedings of the Congress of Evolutionary Computation (CEC 2007), pp. 501–507. IEEE Computer Society Press, Los Alamitos (2007)

Doerr, B., Neumann, F., Sudholt, D., Witt, C.: On the runtime analysis of the 1-ANT ACO algorithm. In: Proceedings of the Genetic and Evolutionary Computation Conference (GECCO 2007), pp. 33–40. ACM Press, New York (2007)

Dorigo, M., Blum, C.: Ant colony optimization theory: A survey. Theoretical Computer Science 344, 243–278 (2005)

Dorigo, M., Stützle, T.: Ant Colony Optimization. MIT Press, Cambridge (2004)

Droste, S., Jansen, T., Wegener, I.: On the analysis of the (1+1) evolutionary algorithm. Theoretical Computer Science 276, 51–81 (2002)

Droste, S., Jansen, T., Wegener, I.: Upper and lower bounds for randomized search heuristics in black-box optimization. Theory of Computing Systems 39(4), 525–544 (2006)

Garnier, J., Kallel, L., Schoenauer, M.: Rigorous hitting times for binary mutations. Evolutionary Computation 7(2), 173–203 (1999)

Giel, O., Wegener, I.: Evolutionary algorithms and the maximum matching problem. In: Alt, H., Habib, M. (eds.) STACS 2003. LNCS, vol. 2607, pp. 415–426. Springer, Heidelberg (2003)

Gutjahr, W.J.: A graph-based ant system and its convergence. Future Generation Computer Systems 16, 873–888 (2000)

Gutjahr, W.J.: A generalized convergence result for the graph-based ant system metaheuristic. Probability in the Engineering and Informational Sciences 17, 545–569 (2003)

Gutjahr, W.J.: Mathematical runtime analysis of ACO algorithms: Survey on an emerging issue. Swarm Intelligence 1, 59–79 (2007)

Gutjahr, W.J.: First steps to the runtime complexity analysis of ant colony optimization. Computers and Operations Research 35(9), 2711–2727 (2008)

Gutjahr, W.J., Sebastiani, G.: Runtime analysis of ant colony optimization with best-so-far reinforcement. Methodology and Computing in Applied Probability 10, 409–433 (2008)

Hart, W.E., Krasnogor, N., Smith, J.E. (eds.): Recent Advances in Memetic Algorithms. Studies in Fuzziness and Soft Computing, vol. 166. Springer, Heidelberg (2004)

Hoos, H.H., Stützle, T.: Stochastic Local Search: Foundations & Applications. Elsevier/Morgan Kaufmann (2004)

Jansen, T., Wegener, I.: Evolutionary algorithms—how to cope with plateaus of constant fitness and when to reject strings of the same fitness. IEEE Transactions on Evolutionary Computation 5, 589–599 (2001)

Kennedy, J., Eberhart, R.C., Shi, Y.: Swarm Intelligence. Morgan Kaufmann, San Francisco (2001)

Krasnogor, N., Smith, J.: A tutorial for competent memetic algorithms: model, taxonomy, and design issues. IEEE Transactions on Evolutionary Computation 9(5), 474–488 (2005)

Levine, J., Ducatelle, F.: Ant colony optimisation and local search for bin packing and cutting stock problems. Journal of the Operational Research Society 55(7), 705–716 (2004)

Lourenço, H.R., Martin, O., Stützle, T.: Iterated local search. In: Handbook of Metaheuristics. International Series in Operations Research & Management Science, vol. 57, pp. 321–353. Kluwer Academic Publishers, Dordrecht (2002)

Merkle, D., Middendorf, M.: Modelling the dynamics of Ant Colony Optimization algorithms. Evolutionary Computation 10(3), 235–262 (2002)

Neumann, F., Wegener, I.: Randomized local search, evolutionary algorithms, and the minimum spanning tree problem. Theoretical Computer Science 378(1), 32–40 (2007)

Neumann, F., Witt, C.: Runtime analysis of a simple ant colony optimization algorithm. In: Asano, T. (ed.) ISAAC 2006. LNCS, vol. 4288, pp. 618–627. Springer, Heidelberg (2006)

Neumann, F., Witt, C.: Ant Colony Optimization and the minimum spanning tree problem. In: Maniezzo, V., Battiti, R., Watson, J.-P. (eds.) LION 2007 II. LNCS, vol. 5313, pp. 153–166. Springer, Heidelberg (2008)

Neumann, F., Sudholt, D., Witt, C.: Rigorous analyses for the combination of ant colony optimization and local search. In: Dorigo, M., Birattari, M., Blum, C., Clerc, M., Stützle, T., Winfield, A.F.T. (eds.) ANTS 2008. LNCS, vol. 5217, pp. 132–143. Springer, Heidelberg (2008)

Neumann, F., Sudholt, D., Witt, C.: Analysis of different MMAS ACO algorithms on unimodal functions and plateaus. Swarm Intelligence 3, 35–68 (2009)

Stützle, T., Hoos, H.H.: MAX-MIN ant system. Journal of Future Generation Computer Systems 16, 889–914 (2000)

Sudholt, D.: Memetic algorithms with variable-depth search to overcome local optima. In: Proceedings of the Genetic and Evolutionary Computation Conference (GECCO 2008), pp. 787–794. ACM Press, New York (2008)

Sudholt, D.: The impact of parametrization in memetic evolutionary algorithms. Theoretical Computer Science 410(26), 2511–2528 (2009)

Wegener, I., Witt, C.: On the optimization of monotone polynomials by simple randomized search heuristics. Combinatorics, Probability and Computing 14, 225–247 (2005)

Witt, C.: Worst-case and average-case approximations by simple randomized search heuristics. In: Diekert, V., Durand, B. (eds.) STACS 2005. LNCS, vol. 3404, pp. 44–56. Springer, Heidelberg (2005)

7

A Multi-resolution GA-PSO Layered Encoding Cascade Optimization Model

Siew Chin Neoh[1], Norhashimah Morad[2], Arjuna Marzuki[1], Chee Peng Lim[1], and Zalina Abdul Aziz[1]

[1] School of Electrical and Electronic Engineering, University of Science Malaysia, Engineering Campus, 14300, Nibong Tebal, Penang, Malaysia
u_jane80@yahoo.co.uk,
{arjuna,cplim,zalina}@eng.usm.my
[2] School of Industrial Technology, University of Science Malaysia, 11800, Minden, Penang, Malaysia
nhashima@usm.my

Abstract. Many real-world problems involve optimization of multi-resolution parameters. In optimization problems, the higher the resolution, the larger the search space, and resolution affects the accuracy and performance of an optimization model. This article presents a genetic algorithm and particle swarm based cascade multi-resolution optimization model, and it is known as GA-PSO LECO. GA and PSO are combined in this research to integrate random as well as directional search to promote global exploration and local exploitation of solutions. The model is developed using the layered encoding representation structure, and is evaluated using two parameter optimization problems, i.e., the Tennessee Eastman chemical process optimization and the MMIC amplifier design interactive optimization.

1 Introduction

Resolution of parameters has significant effects on optimization performance as well as computational requirement. Tuning of parameter settings for higher resolution is difficult due to the requirement of intense search. It is very important to assure a balance between global and local search in order to deal with multi-resolution parameters. To overcome the difficulties associated with multi-resolution problems, much research attention has been focused on the development of multi-resolution representations and approaches.

The foundation for the development of multi-resolution representation and approach is the coarse-to-fine strategy that allows search to be done on different scales. Such strategy promotes the search for global as well as local optimum. The presence of many local minima leads to the difficulty in finding a global minimum. According to [1], the use of the coarse-to-fine strategy is able to prevent convergence into local minima. The optimization approach in [1] starts with the low-resolution approximation of signal (global optimization), passing through the details in its different orientations, and ending with the finest details of the signal (local optimization).

C.P. Lim et al. (Eds.): Innovations in Swarm Intelligence, SCI 248, pp. 121–140.
springerlink.com

According to [2], the adoption of a multi-resolution approach provides optimum management for data representation in which the level of details can be obtained more adequately for a given action. Besides, the multi-resolution approach is claimed to be able to maintain computation under a given bound while improving the accuracy in the proximity of the action focus. For instance, a multi-resolution method to reduce computational cost and enhance the search for the global minimum is proposed in [3], while a multi-resolution analysis to provide sparse storage scheme and reduce both computation time and memory storage is presented in [4].

Multi-resolution representation has been widely used in many applications, especially in the areas of broadcast system and visualization [5-6]. In autonomous aircraft dynamic route optimization, a multi-resolution representation for finding suitable route firstly on a very coarse level, and later on the more detailed levels is used in [7]. On the other hand, Qi and Hunt applied a multi-scale representation to computer verification of handwritten signatures [8]. In artificial intelligence, multi-resolution approaches have been employed as a learning approach in neural networks to improve the performance of signal prediction. In [9], a multi-resolution learning approach is claimed to offer good scalability to complex problem as it provides a systematic hierarchical way to decompose an original learning task and allows exploitation to be accomplished on different correlation structures and resolution levels.

In this research, a GA-PSO LECO model is used to solve parameter optimization. A layered encoding structure is employed as the multi-resolution representation, as explained in Section 2. Details of the proposed GA-PSO-LECO model are presented in Section 3. Applicability of the proposed model to the Tennessee Eastman chemical process and MMIC low noise amplifier design is demonstrated in Sections 4 and 5, respectively. Concluding remarks are presented in Section 6.

2 A Layered Encoding Multi-resolution Representation

An encoding structure is a representation scheme that determines how a problem is structured. Conventional encoding structures for problem representation include bit-string encoding and multidimensional encoding. These methods have a drawback of slow convergence, especially when the resolution increases or when the step of the manipulated variables is small enough to form a large search space.

In this study, a layered encoding structure is used as the solution representation for multi-resolution parameter optimization. A layered encoding structure is different from the existing conventional encoding structures in which it slices the solution representation into multiple layers. Different layers can be used to represent different decisions and objectives, with the aim to ease the analysis and evaluation process. In this research, different layers are employed to represent different resolutions of the parameter value in order to narrow down the search space. The benefit of having a layered encoding structure is that multiple layers promote both global and local search, where an external layer and an internal layer can communicate with each other for its own good (local optimization) and also for the overall advantages (global optimization), updating both global and local optima. Besides, the layered encoding structure is capable of allowing human intervention on the desired layer to expedite the evolutionary search and of incorporating human subjective evaluation for feature

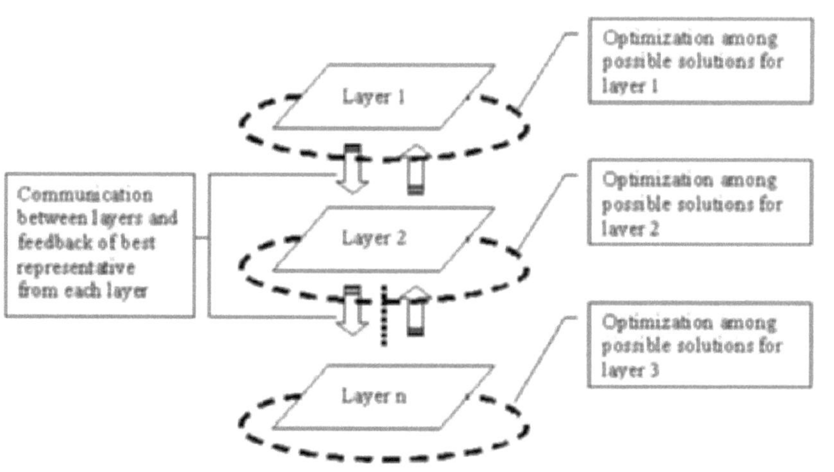

Fig. 1. The layered encoding structure

selection. Fig. 1 shows an example of a 3-layers encoded structure where each layer can play the role as the external or internal layer of each other during the operation of the evolutionary algorithm.

3 The GA-PSO LECO Model

In this research, the Genetic Algorithm (GA) and Particle Swarm Optimization (PSO) models are combined to promote a balance between global and local search. The GA is a stochastic search method inspired by the Darwinian metaphor of natural biological evolution. It is computationally simple, and yet powerful in the search for improvement [10]. It has been widely recognized for its robustness in optimization [11-13]. On the other hand, PSO is a swarm intelligence technique originated by the flocking behavior of birds [14]. In PSO, each potential solution (also known as particle) searches through the problem space, refines its knowledge, adjusts its velocity based on the social information that it gathers, and updates its position. The fittest survival concept of the GA and the swarm flocking perception of PSO motivates the hybridization of the GA and PSO in this research. The GA and PSO hybridization is applied to the layered encoding structure to combine the advantages of randomized and directional search.

The GA search process starts with a randomly initialized population of candidate solutions, known as individuals. These individuals are commonly represented by fixed-length string. In this research, an individual is represented in two layers; in which the external layer indicates the lower resolution representation whereas the internal layer denotes the higher resolution representation. After initialization, these individuals are selected based on fitness to become parents for reproduction of offspring or new individuals. The process of reproduction is carried out using the genetic operators of crossover and mutation. Crossover involves recombination of two preferentially chosen strings (parents) by switching the segments of the strings, keeping some features of both parents in term of genetic materials [15]. As for mutation, it involves alteration of an individual to produce a new solution. The reproduction of new individuals is repeated until the GA termination criterion is met. A pseudo-code description for the GA is as follows.

```
begin
  Let Pc and Pm be the crossover and mutation probability respectively.
  t ← 0;
  Initialize a population of individuals, P(t), with size N at generation t=0;
  Evaluate fitness for each individual;
  while (termination condition is not attained) do
  begin
    Select two parent solutions from P(t) by fitness function;
    Randomly generate a number, r, 0≤r≤1;
    if(r<Pc)
        Carry out crossover operation on the selected solutions to produce child solutions;
    end
    Randomly generate a number, r, 0≤r≤1;
    if(r<Pm)
        Carry out mutation operation on the selected solution to produce child solution;
    end
    Evaluate fitness of child solutions
    Reinsert the child solutions in the new population based on fitness;
    t ← t+1;
  end
end
```

PSO is an evolutionary technique that does not use filtering operations such as crossover and mutation. The algorithm starts with the initialization of candidate solutions called particles. The associated position and velocity of each particle are randomly generated. Then, using the fitness evaluation function, compare the *pbest* (best found position or coordinate for a particle) fitness of each particle to its current fitness. If the current fitness is better, then update the *pbest* fitness to the current fitness, and assign the current position to replace the *pbest* position or coordinate. Similarly, compare the fitness of each particle with the *gbest* (best found position for all particles) fitness. If the current fitness is better than *gbest*, replace *gbest* with the current position, and assign the current fitness as the *gbest*'s fitness. The associated position and velocity for each particle are updated based:

$$vel_{ikq_m} = w \times vel_{ikq_{m-1}} + c_1 \times rand(\ ) \times (pbest - Pos_{ikq_{m-1}}) + c_2 \times rand(\ ) \times (gbest - Pos_{ikq_{m-1}}) \quad (1)$$

$$Pos_{ikq_m} = Pos_{ikq_{m-1}} + vel_{ikq_m} \quad (2)$$

where vel_{ikq} and Pos_{ikq} refer to the velocity and position of particle q in the search space (coordinate) of $i \times k$, respectively. m is the iteration number, w is the inertia weight, c_1 and c_2 are two constant factors called the cognitive parameter and social parameter, respectively, and rand( ) is a randomly generated value between 0 and 1.

Particle positions and velocities are refined repeatedly from iteration to iteration to direct the candidate solutions towards the optimum position until the termination criterion is met. A pseudo-code description for PSO is as follows.ï

```
begin
  Let iter denote the iteration number of PSO.
  iter ← 0;
  Initialize a population of particles, Pop(iter), with size n at iter=0;
  while (termination condition is not attained) do
  begin
    for each particle
    Calculate fitness value;
      if the fitness value is better than the best fitness value (pbest) in history
        Set current value as the new pbest
      end
    end
    Choose the particle with the best fitness value of all the particles as the gbest
      for each particle
        Calculate particle velocity according equation (2)
        Update particle position according equation (3)
      end
  end
end
```

In this research, the GA and PSO are combined with the proposed layered encoding structure to form a hybrid GA-PSO LECO model. This model integrates the randomized feature of the GA to encourage exploration for global solutions, and, at the same time, incorporates the swarm learning behavior of PSO to motivate the exploitation of local area solutions through directional search. The underlying idea of GA-PSO LECO is to ensure a balance between exploration and exploitation of search.

The layered encoding structure that is built up by multiple layers stimulates the development of a cascade evolutionary model, whereby optimization can be firstly focused on a particular layer (external layer) and then followed by the others (internal layers). For each particular external layer, there could be thousands of possible internal layers. In such case, a cascade flow of evolutionary search is delivered. Fig. 2 shows the general flow of the developed multi-resolution GA-PSO LECO model. In the model, N individuals that represent the first layer parameter setting in integer values is randomly initialized.

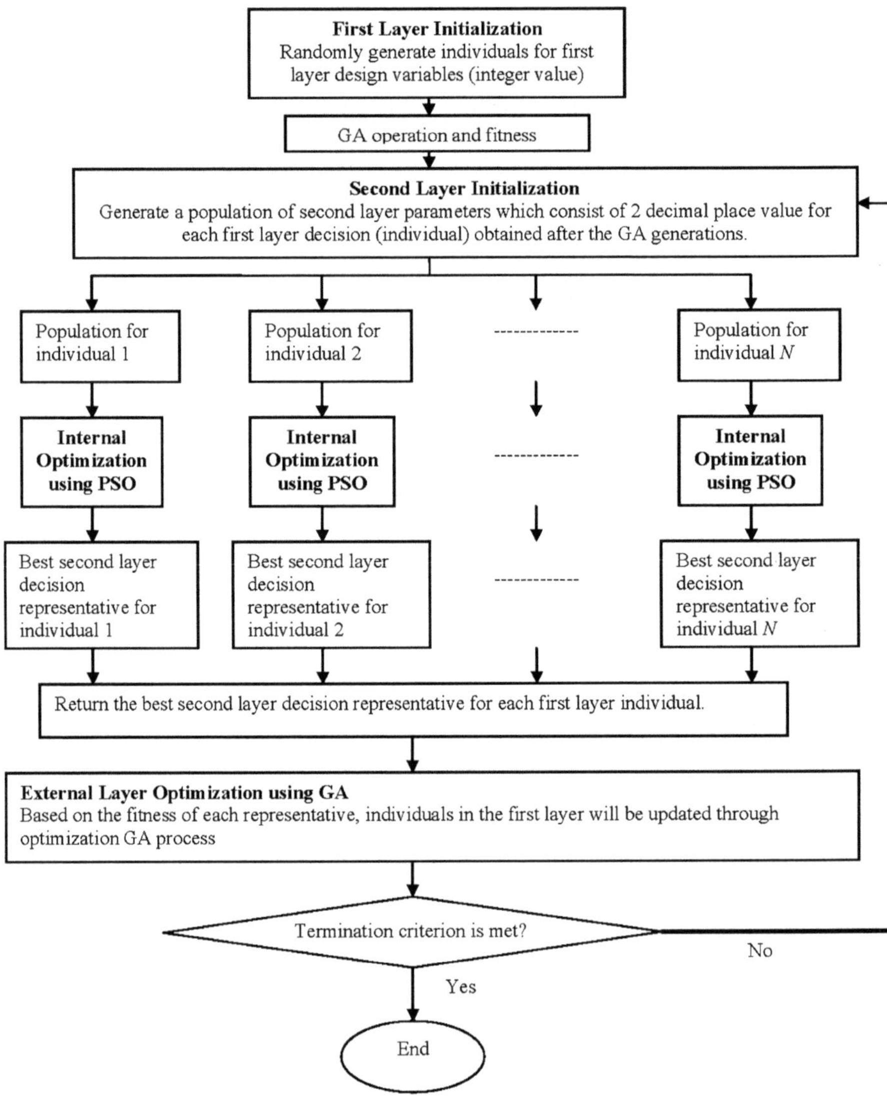

Fig. 2. General flow of the developed multi-resolution GA-PSO LECO model

The GA is firstly applied to roughly approximate good parameter setting in integer values. Then, for each individual, a population of the second layer parameter setting in higher resolution values is generated. PSO is used as the internal layer optimizer to find the best second layer representative for each first layer individual. The best second layer representative combined with the first layer individual then undergoes the GA optimization process to fine-tune the first layer parameter setting. If the termination criterion is not met, the flow is repeated again in which a new population of the second layer parameter setting solutions is initialized for each updated first layer parameter setting solution. The GA-PSO LECO model in this research is developed in Matlab. A pseudo-code description for the GA-PSO hybrid operation is as follows.

```
begin
  GA operation for lower resolution parameter setting;
  while (termination condition is not attained) do
  begin
    for each selected individual of GA
      PSO operation for higher resolution parameter setting;
    end
    return the best second layer representative (higher resolution setting) to each
first layer individual (lower resolution setting)
    begin
        GA operation for lower resolution parameter setting (update first layer indi-
        viduals)
    end
  end
end
```

4 Tennessee Eastman Chemical Process Optimization

The Tennessee Eastman (TE) chemical process is a plant-wide industrial control problem proposed in [16]. It is a large scale, continuous, and nonlinear process [17]. In this research, the TE process is adopted for the study on multi-resolution parameter optimization. There are five major operation units in the TE process: a reactor, a product condenser, a vapor-liquid separator, a stripper, and a compressor. The detailed process flow is available in [16]. The TE problem used in this research is downloaded from http://depts.washington.edu/control/LARRY/TE/download.html. From there, the simulation model of the control strategy described in [18] is used to study the effectiveness of the GA-PSO LECO model. The simulation model is depicted in Fig. 3.

Table 1 shows the specific operational constraints in the TE process. These constraints are important for equipment protection. When the specific process variables are out of control, the process will be shut down. In [19], these specific variables are among the various set points (decision variables) used to optimize the operating cost function. Considering the specific operational constraints listed in Table 1 and the set points given by [19], the reactor level, stripper level, separator level, product flow, reactor pressure, and reactor temperature are used as the parameters for optimizing the production operating cost in this research. The parameter ranges used are given in Table 2.

The total operating cost is the objective function given in [16]. It has been used as the objective function for optimization in [19] and [20]. The total operating cost is calculated in cost per hour ($/h), considering reactant and product losses in the purge and product streams, stripper steam use, and compressor work. Equation (3) shows the total operating cost function.

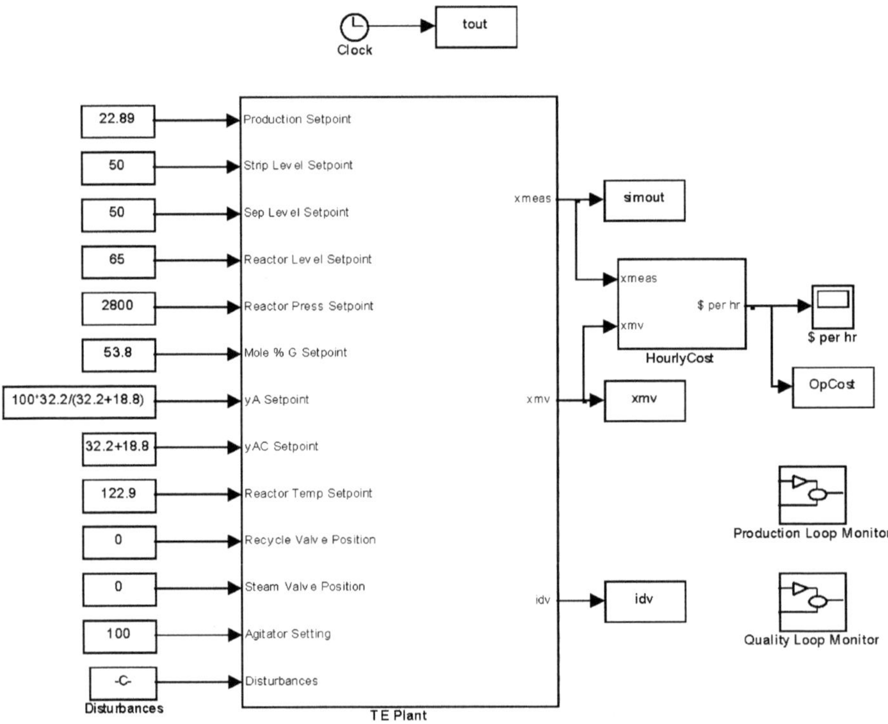

Fig. 3. A decentralized control model of the Tennessee Eastman Process developed by Ricker (1996)

Table 1. Process operating constraints obtained from Downs and Vogel (1993)

Process Variable	Normal Operating Limits		Shut Down Limits	
	Low Limit	High Limit	Low Limit	High Limit
Reactor Pressure	none	2895 kPa	None	3000 kPa
Reactor Level	50% (11.8 m^3)	100% (21.3 m^3)	2.0 m^3	24.0 m^3
Reactor Temperature	none	150 °C	None	175 °C
Product Separator Level	30% (3.3 m^3)	100% (9.0 m^3)	1.0 m^3	12.0 m^3
Stripper Base Level	30% (3.5 m^3)	100% (6.6 m^3)	1.0 m^3	8.0 m^3

Table 2. Parameter ranges used for the TE problem in this study

Process variable (Parameter)	Set point range	
	Low Limit	High Limit
Reactor Pressure	2600	2850 kPa
Reactor Level	50% (11.8 m^3)	100% (21.3 m^3)
Reactor Temperature	120 ºC	150 ºC
Product Separator Level	30% (3.3 m^3)	100% (9.0 m^3)
Stripper Base Level	30% (3.5 m^3)	100% (6.6 m^3)
Product Flow	20 m^3h^{-1}	25 m^3h^{-1}

$$C_{tot.} = F_9 \sum_{i=A,i\neq B}^{H} C_{i,cst} X_{i,9} + F_{11} \sum_{i=D}^{F} C_{i,cst} X_{i,11} + 0.0536 W_{cmp} + 0.0318 F_{steam} \quad (3)$$

The cost of each component, $C_{i,cst}$ which is calculated in $\$kg^{-1}mol^{-1}$ is provided by [16]. W_{cmp} and F_{steam} refer to the compressor power and steam flow rate, whereas F_9, $X_{i,,9}$, and F_{11}, $X_{i,11}$ refer to the molar flow rate and mole fraction of component '*i*' in streams 9 and 11, respectively. As the TE simulation is a continuous process, the average total operating cost of the whole simulation duration is taken as the objective function in this research.

4.1 A Layered Encoding Structure for the TE Process

Resolution that is able to provide a more detailed level of information is highly related to accuracy of the search [2]. In TE parameter optimization, resolution refers to the decimal place in parameter values. Using the layered encoding structure, decisions for the parameter value set are represented in multiple layers or separated strings. Each layer represents different resolution values of the parameters. Fig. 4 presents the layered encoding structure for representing parameter values with different resolutions. In the structure, two layers are used. The first layer represents the parameter values in integer whereas the second layer represents the parameter values in two decimal place resolution. The coarser resolution level in the first layer and the finer resolution level in the second layer promote search for global and local optima. A mechanism can be applied to the particular resolution level for optimization using the multi-resolution layered encoding structure. In this study, the coarser resolution level is optimized using the GA whereas the finer resolution level is optimized using PSO.

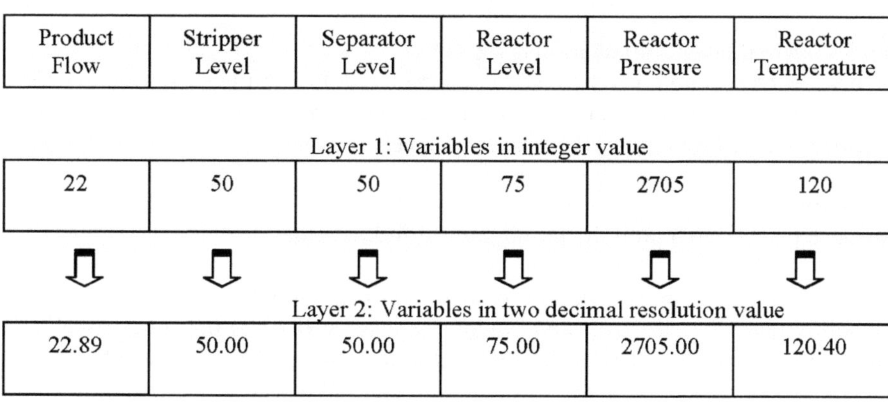

Fig. 4. A layered encoding structure for representing parameter resolution in TE

4.2 The GA-PSO LECO Model for TE Parameter Optimization

The GA-PSO LECO model is developed to optimize the parameters in the TE process. The aim is to minimize the mean operating cost throughout the process duration. Using the layered encoding multi-resolution representation, optimization is first carried out on separate layers so as to have an in depth local search. Then, the layers are combined for a global search. The proposed GA-PSO LECO model uses global updates to allow the search to escape from local traps. By referring to Fig. 2, the detailed procedure of the GA-PSO LECO model in optimizing the TE process is as follows.

Step 1: First layer initialization
Randomly generate a population of first layer individuals (parameter setting in integer values)

Step 2: GA operation
Obtain a rough solution position in the coarser resolution level using the GA operation. This step provides an approximation for the solution before going into a detailed local search. A coarse resolution approximation is intended to reduce the computation effort so that only a sparse computation space is considered at a time.

Step 3: Second layer initialization
Based on the approximated population of first layer individuals, a population of second layer parameter settings in two decimal place resolution is generated for each first layer individual in accordance with the coarse level setting.

Step 4: Internal layer (second layer) optimization using PSO
PSO is used to find the best second layer representative for each first layer individual. The aim is to focus on the in-depth finer level resolution search so as to find the finer resolution setting that best fits the coarse resolution setting in each first layer individual.

Step 5: External layer optimization using GA
The best found second layer setting is combined with the first layer setting to undergo an external layer optimization process using the GA. In this operation, global updates are carried out to fine-tune the first layer parameter settings.

Step 6: Repeat
Repeat step 3 to step 5 until the termination criterion is met.

In the GA operation, two-point crossover and single-point mutation are used. The probability rates for crossover and mutation are set at 0.7 and 0.3, respectively. The population size is 10, whereas the maximum generation is 5. As for PSO, the decreasing inertia weight, w, is set from the range of 0.9 to 0.4, and constant factors, c_1 and c_2 are set to 2. The population size and the maximum iteration for PSO are both set at 5. The termination criterion of GA-PSO LECO is reached when the whole GA-PSO flow is repeated for 4 times.

4.3 Results and Discussion

Table 3 presents a comparison between the best parameter settings of GA-PSO LECO and Ricker's decentralized model (obtained from http://depts.washington.edu/control/LARRY/TE/download.html). From Table 3, it is shown that the mean total operating cost obtained by GA-PSO LECO (104.1835 per hour) is less than that of Ricker's model ($120.0238 per hour). The difference between the two models (i.e., $16 per hour) may not be much. But, when the optimization process is carried out in a longer duration, the amount of difference in total operating cost is expected to be highly increased.

In order to assess the stability of GA-PSO LECO towards optimizing the TE parameter, 30 repetition runs are carried out. The mean of the total operating cost ($/h) is shown in Fig. 5. From Fig. 5, it is clear that the maximum value of the mean total operating cost obtained by GA-PSO LECO is less than $117 per hour, whereas the minimum total operating cost is about $104 per hour. The repeated results between $104 per hour and $117 per hour show the capability of GA-PSO LECO in finding better parameter settings that can improve the results from Ricker's model.

Table 3. Comparison between the GA-PSO LECO model and Ricker's decentralized model

TE Parameter Setting	GA-PSO LECO Model	Ricker Decentralized Test Model
Product Flow (m^3/h)	22.00	22.89
Stripper Level (%)	30.00	50
Separator Level (%)	62.00	50
Reactor Level (%)	50.00	65
Reactor Pressure (kPa)	2789.00	2800
Reactor Temperature (°C)	122.00	122.9
mean total operating costs ($/h)	104.1835	120.0238

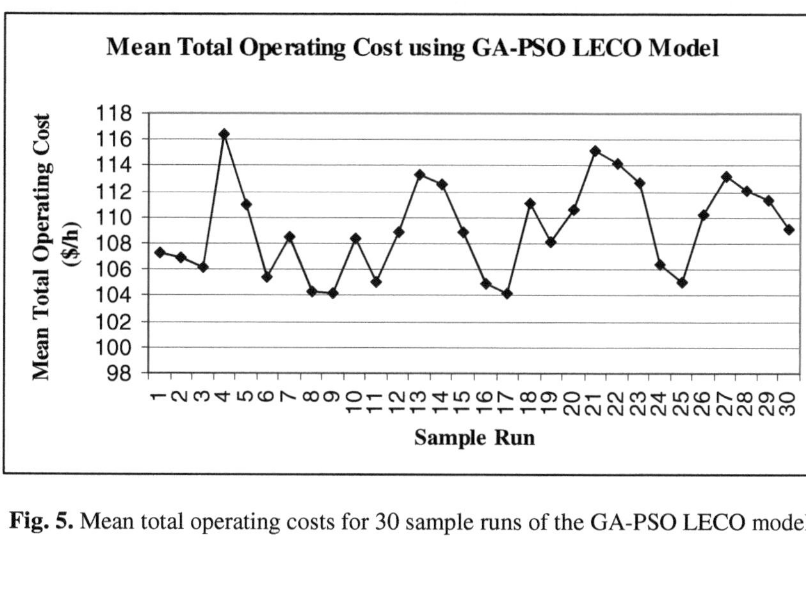

Fig. 5. Mean total operating costs for 30 sample runs of the GA-PSO LECO model

5 Interactive MMIC Low Noise Amplifier Design Optimization

Broadband amplifier is in demand for broadband wireless application such as ultra wideband application [21] and multiple-band application such as Personal Communication Service (PCS), International Mobile Telecomunication-2000 (IMT-2000) and wireless local area network (WLAN). With the increasing demand of low noise monolithic RF receivers for wireless communication, particular interests have been paid to the design of a Low Noise Amplifier (LNA) [22-23].

In the design of LNA, low noise figure, high gain, and low power consumption are crucial parameters. In this research, an RC feedback topology is used as the architecture of the MMIC broadband amplifier. The MMIC circuit design is devised using the Agilent Advance Design System (ADS) [24]. The ADS is a circuit design software that comprises a number of built-in optimizers such as Random Optimizer, Gradient Optimizer, and Quasi-Newton. The stand-alone characteristic of the ADS, however, makes the integration of external optimizers difficult because a large expenditure is required for training and time spent on translation when switching the programs to allow interoperability. As a result, an interactive MMIC low noise amplifier circuit design optimization process that integrates human intervention in computation is carried out.

This case study involves 12 design variables. The aim is to optimize the power gain, dB(S(2,1)), noise figure, NFmin_out, and drain current, ID. Besides optimization, it is aimed to fulfill the geometric stability factors such as mu_source and mu_load. Fig. 6 shows the RC feedback amplifier schematic which consists of two stages of RC feedback amplifiers. M_1 and M_2 are transistors and the unit gate width (UGW) of M_1 and M_2 is 50μm. C_{CC} in the circuit is a coupling capacitor while L, C_C, R_S, R_L are external components with R_S and R_L equal to 50 Ω, C_C equals to 100 pF, and L equals to 0.1 μH.

Input resistor, R_{IN}, input capacitor, C_{IN}, output resistor, R_{OUT}, output capacitor, C_{OUT}, output resistor of the second stage amplifier, R_{OUT2}, and output capacitor of the second stage, C_{OUT2}, are primarily used to stabilize the circuit whereas source resistor, R_S, bypass capacitor, C_S, source resistor of the second stage, R_{S2}, bypass capacitor of

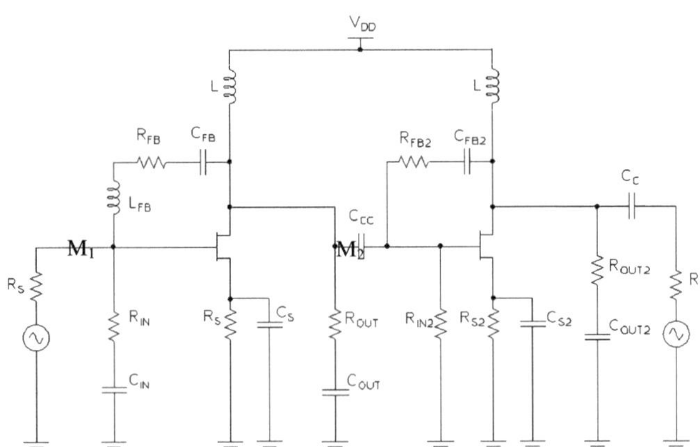

Fig. 6. An RC feedback amplifier circuit

the second stage, C_{S2}, and input resistor of the second stage, R_{IN2}, are used to bias M_1 and M_2. This configuration ensures voltage at gate M_1 and M_2 to be 0 V. In addition, feedback resistor, R_{FB}, feedback capacitor, C_{FB}, feedback inductor, L_{FB}, feedback resistor of the second stage amplifier, R_{FB2} and feedback capacitor of the second stage, C_{FB2}, are used as feedback networks for M_1 and M_2, respectively. These components undoubtedly affect the performance of the whole circuit.

Table 4 shows the design variables that are considered for amplifier optimization. As device fingers contribute to different power gain throughout the operating frequency, the transistor finger value is also considered as one of the design variables. These variables are manipulated to achieve the required specifications and objectives as shown in Table 5. The intended optimization scheme and weight for each output parameter are based on the original MMIC circuit design as in [24].

Table 4. Synthesis setup of design variables

Design Variables	From	To
R_{FB} (Ohm)	10	550
Finger	2	10
C_{FB} (pF)	4	7
R_{OUT} (Ohm)	100	300
C_{OUT} (pF)	2	20
R_{IN} (Ohm)	500	1000
C_{IN} (pF)	2	20
L_{FB} (nH)	1	10
C_{CC} (pF)	1	20
R_S (Ohm)	5	50
C_S (pF)	1	20
R_{S2} (Ohm)	5	50

Table 5. The required specifications of amplifier optimization

Output Parameters	Value Range		Optimization	Weight
	Minimum	maximum		
NFmin_out (dB)	-----	2.5	Minimize	5
dB(S(2,1))	17	20	Maximize	1
mu_load	1.05	-----	-----	1
mu_source	1.05	-----	-----	1
ID (A)	0.04	0.05	Minimize	10

The built-in optimizers in the ADS, i.e., Random Optimizer, Gradient Optimizer, and Quasi-Newton are applied to optimize the design variables of the MMIC amplifier. The results obtained from each built-in optimizer are compared with those from GA-PSO LECO. Owing to the difficulty in incorporating the Matlab software into the ADS, the genetic search operation for this case study is conducted offline.

5.1 The Layered Encoding Structure for MMIC Low Noise Amplifier Parameter Optimization

Multi-resolution parameter optimization is computationally intractable especially when optimization search is carried offline whereby the parameter or variable values need to be manually inserted to an external design or simulation. The layered encoding structure is different from conventional encoding in a way that it promotes both global and local search where the external and internal layers can communicate to update both global and local optima. Besides, it permits human intervention where the user can select the most preferred external or internal layer for chromosome masking. Furthermore, the layers could enhance visualization because mapping of n-dimensional solution to 2-dimensional solution is not required as the layered encoding structure by itself is a straightforward 2-dimensional representation. These advantages give the layered encoding structure an easy access to analyze and retrieve particular information from the problem representation.

A two-layered encoding structure is used as the solution representation for amplifier design, as shown in Fig. 7. Different slices of layer represent different resolution values of the design variables. The external layer presents the design variables in integer values whereas the internal layer shows 3-decimal place resolution values for the designated variables. This representation aims to narrow down the search space, allowing an in-depth search for local optimum based on a coarser resolution level. After finding the local optimum, multiple layers are combined to obtain the global optimum.

5.2 Interactive GA-PSO LECO Approach

Online optimization allows the evolutionary search to directly optimize a particular design. However, many real-world applications face issues on system and software incompatibility. When optimization is conducted offline, human fatigue becomes a serious factor. As a result, optimization based on Interactive Evolutionary Computation [IEC], which is able to accelerate the genetic search process and, at the same time, to reduce human fatigue, is investigated.

Most IEC models use human evaluation in fitness definition and chromosome masking [25-28]. In this case study, human evaluation is applied to selection of the number of individuals for reproduction. Human intervention aims to narrow down the solution search space in order to accelerate EC convergence, and, at the same time, reduce human fatigue. Besides, it also prevents some potential solutions from being discarded owing to some bias fitness value.

In this research, human intervention is first incorporated in the GA selection and reproduction process. Then, feature extraction is done by the layer masking procedure. By referring to Fig. 2, the detailed procedure of the interactive GA-PSO LECO model used for optimizing the MMIC circuit design is as follows.

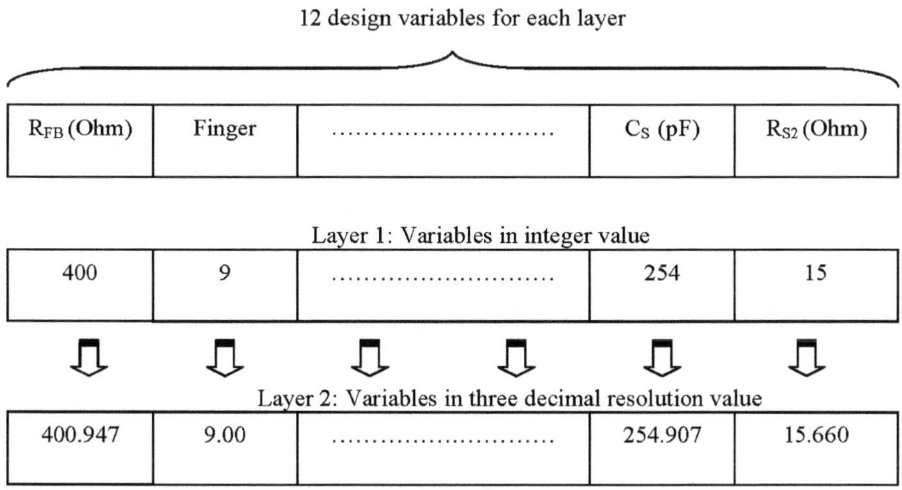

Fig. 7. The layered representation of design variables in amplifier optimization

Step1: First (External) Layer Initialization
In the first layer (external layer) initialization, design variables for the MMIC amplifier are randomly initialized in integer values based on the design variable's specifications, as given in Table 4. The GA (population size=50, generation=4) is then used to optimize the external layer to obtain the fitness value for evaluation.

Step2: Human Intervention
Based on the fitness of each solution, human intervention is introduced to select and decide the number of the first layer individuals to undergo an in-depth second layer optimization as well as to be reproduced for elitism.

Step3: Second (Internal) Layer Initialization
The second layer decisions with a higher resolution (3-decimal place) design variable value are generated for each selected first layer individual. PSO (particle size=10, iteration=3) is integrated to find the best second layer decision representation for each of the first layer individual that has been selected by human evaluation in Step 2.

Step 4: External Layer Updating
By referring to the internal layer representative, the fitness of the first layer decision is updated to undergo a further GA process in which a new generation of the first layer individuals is generated for human evaluation.

Step 5: Repeat
Repeat step 2 to step 4 until the termination criterion is met. In this case, the termination criterion is based on human subjective evaluation on the objective fitness and the specification requirements.

In order to reduce human fatigue, human intervention is not applied to every GA generation. Therefore, in the normal GA operation, stochastic universal sampling is used as the selection function. Four-point crossover and two-point mutation are used whereby the crossover probability is set at 0.8 and the mutation probability is at 0.2. On the other hand, PSO is conducted with an inertia weight ranges from 0.4 to 0.9, and with the two constant factors, c_1 and c_2, set to 2. The genetic and particle swarm parameters are selected based on the results after several trials.

In fitness evaluation, a weighted-sum approach is used because preference or bias is intended for the amplifier design. The weights used for the specified criteria (Table 5) are the same as those in [24]. In order to have a fair comparison, all other optimization techniques employ the weighted-sum approach with the same weights too. Specified criteria that do not meet the minimum or maximum constraint as shown in Table 5 are given a penalty value of 10. For Eq (4-8), *F1*, *F2*, *F3*, *F4*, and *F5* are fitness for noise figure (NFmin_out), power gain (dB(S(2,1))), mu_load, mu_source, and current (ID) respectively.

$$F1 = \begin{cases} 10 + (\mathrm{NF_{min\_out}} - 2.5), \text{if NFmin\_out} > 2.5 \\ 10 + (0.1 - \mathrm{NF_{min\_out}}), \text{if NFmin\_out} < 0.1 \\ \mathrm{NF_{min\_out}} - 0.1, \text{if } 0.1 < \text{NFmin\_out} < 2.5 \end{cases} \tag{4}$$

$$F2 = \begin{cases} 10 + (\mathrm{dB(S(2,1))} - 20), \text{if dB(S(2,1))} > 20 \\ 10 + (17 - \mathrm{dB(S(2,1))}), \text{if dB(S(2,1))} < 17 \\ 20 - \mathrm{dB(S(2,1))}, \text{if } 17 < \mathrm{dB(S(2,1))} < 20 \end{cases} \tag{5}$$

$$F3 = \begin{cases} 10, \text{if mu\_load} < 1.05 \\ 0, \text{if mu\_load} > 1.05 \end{cases} \tag{6}$$

$$F4 = \begin{cases} 10, \text{if mu\_source} < 1.05 \\ 0, \text{if mu\_source} > 1.05 \end{cases} \tag{7}$$

$$F5 = \begin{cases} 10 + (0.04 - \mathrm{ID}), \text{if ID} < 0.04 \\ 10 + (\mathrm{ID} - 0.05), \text{if ID} > 0.05 \\ \mathrm{ID} - 0.04, \text{if } 0.04 < \mathrm{ID} < 0.05 \end{cases} \tag{8}$$

The objective normalization method used in this research is taken from [29]. The total objective value for the evaluation function is calculated based on the relative fitness of the individuals with respect to the same objective as in Eq. (9).

$$F_{tot} = \lambda_j \sum_{j=1}^{ob} \frac{F_j}{F_{average}} \tag{9}$$

where,
$F_{average}$=the average objective value of objective j at generation 1
F_j=objective value of each individual for objective j
F_{tot}=total objective value
λ_j=weight for objective j

Considering Eq. (9) and the objective weight given in Table 5, the total summation of all objective values, F_{tot}, is derived as in Eq. (10). The main objective in optimizing the MMIC amplifier design variables is to reduce F_{tot} at the frequency of 3GHz.

$$F_{tot} = 5\frac{F1}{F1_{average}} + \frac{F2}{F2_{average}} + \frac{F3}{F3_{average}} + \frac{F4}{F4_{average}} + 10\frac{F5}{F5_{average}} \quad (10)$$

5.3 Results and Discussion

The optimized variables for the MMIC amplifier design obtained from GA-PSO LECO are given in Table 6. The value of each specified parameter (objective) and the total fitness value obtained from different optimizers are shown in Table 7.

Table 6. Optimized design variables for the MMIC amplifier using the interactive GA-PSO LECO model

Design Variable	Optimized Value given by Interactive GA-PSO LECO model
R_{FB} (Ohm)	72.599
Finger	7.000
C_{FB} (pF)	6.091
R_{OUT} (Ohm)	247.506
C_{OUT} (pF)	12.883
R_{IN} (Ohm)	912.615
C_{IN} (pF)	5.046
L_{FB} (nH)	10.951
C_{CC} (pF)	19.169
R_S (Ohm)	41.826
C_S (pF)	37.611
R_{S2} (Ohm)	30.846

Table 7. Parameter Value of Different Optimizer

Optimizer	Nfmin_out	dB(S(2,1))	mu_load	mu_source	ID	Ftot
Gradient	2.216	19.829	5.26	2.734	0.0499619	2.33511
Random	2.216	19.315	6.82	2.9	0.0493191	2.37448
Quasi-Newton	2.267	19.45	5.43	2.692	0.0493913	2.41968
Interactive GA-PSO LECO	1.425	17.838	1.109	1.087	0.0466944	1.62338

Referring to Table 7, all optimizers including GA-PSO LECO produce a value higher than 1 for mu_load and mu_source, i.e., the designed amplifier is stable. Besides, it is observed that GA-PSO LECO outperforms other optimizers in minimizing noise figure (NFmin_out) and current (ID). Although GA-PSO LECO yields a lower power gain, i.e., dB(S(2,1)), it should be noted that NFmin_out and ID are more important parameters than dB(S(2,1). As a result, GA-PSO LECO produces a better amplifier design in the overall performance. This can be further ascertained by referring to the total fitness value whereby GA-PSO LECO gives the lowest F_{tot}, as compared with those from other optimizers. In summary, GA-PSO LECO is a flexible tool that is capable of integrating a multi-resolution structure for parameter optimization as well as of handling human intervention.

6 Summary

A novel GA-PSO LECO model that combines the advantages of hybrid GA-PSO evolutionary search as well as the layered encoding representation structure has been demonstrated to be an effective approach for undertaking multi-resolution optimization tasks. The proposed model has been evaluated using the TE chemical process and the MMIC low noise amplifier design problems. The results indicate that GA-PSO LECO is able to achieve good performances in multi-resolution parameter optimization. In the TE problem, the parameter setting proposed by GA-PSO LECO is able to give a better total operating cost, as compared with that of Ricker's model. As for the MMIC amplifier design, the GA-PSO-LECO model has shown to be an alternative that is able to outperform other optimizers such as Random, Gradient, and Quasi-Newton. In addition, the proposed model is a flexible tool that allows human intervention to realize an interactive evolutionary search mechanism for solving issue related to software incompatibility besides multi-resolution problems.

References

1. Gefen, S., Tretiak, O., Bertrand, L., Rosen, G.D., Nissanov, J.: Surface alignment of an elastic body using a multi-resolution wavelet representation. IEEE Transactions on Biomedical Engineering 51(7), 1230–1241 (2004)
2. Ganovelli, F., Cignoni, P., Montani, C., Scopigno, R.: Enabling cuts on multiresolution representation. In: IEEE Proceedings of Computer Graphics International, pp. 183–191 (2000)
3. Law, Y.N., Lee, H.K., Yip, A.M.: A multi-resolution stochastic level set method for the Mumford-Shah segmentation of bioimages. In: 8th World Congress on Computational Mechanics (2008)
4. Loison, R., Gillard, R., Citerne, J., Piton, G., Legay, H.: Optimised 2D multi-resolution method of moment for printed antenna array modeling. IEE proceedings of Microwave, Antennas Propagation 148(1), 1–8 (2001)
5. Uhercik, M., Kybic, J., Liebgott, H., Cachard, C.: Multi-resolution parallel integral projection for fast localization of a straight electrode in 3D ultrasound images. In: 5th IEEE International Symposium on Biomedical Imaging, Paris, pp. 33–36 (2008)

6. Sehlstedt, M., LeBlanc, J.P.: A computability strategy for optimization of multiresolution broadcast systems: a layered energy distribution approach. IEEE Transactions on Broadcasting 52(1), 11–20 (2006)
7. Samad, T., Gorinevsky, D., Stoffelen, F.: Dynamic multiresolution route optimization for autonomous aircraft. In: Proceedings of the IEEE 2001 International Symposium on Intelligent Control, Mexico, pp. 13–18 (2001)
8. Qi, Y.Y., Hunt, B.R.: A multiresolution approach to computer verification of handwritten signatures. IEEE Transactions on Image Processing 4(6), 870–874 (1995)
9. Liang, Y., Liang, X.: Improving signal prediction performance of neural networks through multiresolution learning approach. IEEE Transactions on Systems, Man, and Cybernetics-Part B: Cybernetics 36(2), 341–352 (2006)
10. Goldberg, D.E.: Genetic algorithms: In search, optimization and machine learning. Addison-Wesley, USA (1989)
11. Fogel, D.: Evolutionary Computation: Toward a new philosophy of machine intelligence. IEEE Press, Piscataway (1995)
12. Shin, K., Lee, Y.: A genetic algorithm application in bankruptcy prediction modeling. Experts Systems with Applications 23(3), 321–328 (2002)
13. Gen, M., Cheng, R.: Genetic algorithms and engineering optimization. John Wiley & Sons Inc., New York (2000)
14. Kennedy, J., Eberhart, R.C.: PSO optimization. In: Proceedings of the IEEE International Conference on Neural Networks, vol. 4, pp. 1942–1948 (1995)
15. Padhy, N.P.: Artificial intelligence and intelligent systems. Oxford University Press, India (2005)
16. Downs, J.J., Vogel, E.F.: A plant-wide industrial process control problem. Computers Chemical Engineering 17(3), 245–255 (1993)
17. Yan, M., Ricker, N.L.: Multi-objective control of the Tennessee Eastman challenge process. In: Proceedings of the American Control Conferences, Seattle, Washington, pp. 245–249 (1995)
18. Ricker, N.L.: Decentralized control of the Tennessee Eastman challenge process. Journal of Process Control 6(4), 205–221 (1996)
19. Golshan, M., Boozarjomehry, R.B., Pishvaie, M.R.: A new approach to real time optimization of the Tennessee Eastman challenge problem. Chemical Engineering Journal 112, 33–44 (2005)
20. Duvall, P.M., Riggs, J.B.: Online optimization of the Tennessee Eastman challenge problem. Journal of Process Control 10, 19–33 (2000)
21. Bevilacqua, A., Niknejad, A.M.: An Ultra-Wideband CMOS LNA for 3.1 to 10.6 GHz wireless receiver. In: IEEE Int. Solid-State Circuits Conference, San Francisco, vol. 1, pp. 382–533 (2004)
22. Belostotski, L., Haslett, J.W.: Noise figure optimization of inductively degenerated CMOS LNAs with integrated gate inductors. IEEE transactions on Circuits and Systems I 53(7), 1409–1422 (2006)
23. An, D., Rhee, E.-H., Rhee, J.-K., Kim, S.D.: Design and fabrication of a wideband MMIC low-noise amplifier using Q-Matching. Journal of the Korean Physical Society 37(6), 837–841 (2000)
24. Marzuki, A., Sauli, Z., Md Shakaff, A.Y.: A practical high frequency integrated circuit power-constraint design methodology using simulation-based optimization. In: United Kingdom-Malaysia Engineering Conference, London (2008)

25. Nishio, K., Murakami, M., Mizutani, E., Honda, N.: Fuzzy fitness assignment in an interactive genetic algorithm for a cartoon face search. In: Genetic Algorithm and Fuzzy Logic Systems, Soft Computing Perspectives. Advances in Fuzzy Systems Applications and Theory, vol. 7, pp. 175–192 (1997)
26. Caldwell, C., Johnston, V.S.: Tracking a criminal suspect through 'Face-Space' with a genetic algorithm. In: Proc. 4th Int. Conf. Genetic Algorithms, pp. 416–421. Morgan Kaufman, San Diego (1991)
27. Smith, J.R.: Designing biomorphs with an interactive genetic algorithm. In: Proc. 4th Int. Conf. Genetic Algorithms, San Diego, pp. 535–538 (1991)
28. Hsu, F.C., Chen, J.-S.: A study on multi criteria decision making model: interactive genetic algorithms approach. In: Proc. IEEE Int. Conf. on System, Man, and Cybernetics, Tokyo, Japan, pp. 634–639 (1999)
29. Morad, N.: Optimization of Cellular Manufacturing Systems Using Genetic Algorithms, Ph.D. Thesis. University of Sheffield, Sheffield, UK (1997)

8

Integrating Swarm Intelligent Algorithms for Translation Initiation Sites Prediction

Jia Zeng and Reda Alhajj

Department of Computer Science, University of Calgary, Calgary, Alberta, Canada
Department of Computer Science, Global University, Beirut, Lebanon
jzeng@ucalgary.ca,alhajj@ucalgary.ca

Abstract. Translational initiation sites (TISs) are an important type of gene signals which flag the starting location of the translation process. Due to the characteristics of the translational mechanism, an accurate recognition of TIS in a messenger RNA sequence leads to the determination of the primary structure of the corresponding protein. Many existing TIS prediction approaches investigate the data from one single perspective or apply some static central fusion mechanism on a fixed set of features. Due to the complicated nature of the genetic data, we believe that it is beneficial to consider multiple biological perspectives in the analysis process. In order to provide diversified problem solving techniques as well as modularization to the system, we have proposed a novel solution that uses a multi-agent system (MAS), which investigates the biological data from multiple biological perspectives, each of which is implemented by an independent problem solver agent with a unique expertise. A generalized layered framework is proposed to facilitate the system to take advantage of the synergy of having multiple agents working together and arriving at a single final prediction. In this chapter, we explore the application of particle swarm optimization and ant colony optimization in the proposed architecture. The integration of the swarm intelligent algorithms has lead to an outstanding performance of the system. Extensive experiments on three benchmark data sets have verified this claim, demonstrating the advantage of using the proposed system over most of the existing TIS prediction approaches.

1 Introduction

Swarm intelligence is a branch of artificial intelligence that develop algorithms inspired by the emergent behavior observed in a large number of small organisms such as flocks, herds, fish schools, ant colony, etc. Examples of some well-known swarm intelligence include the ant colony optimization, particle swarm optimization, and stochastic diffusion search. In this chapter, we will investigate the effectiveness of applying swarm intelligence to predict translational initiation sites in various types of nucleotide sequences.

C.P. Lim et al. (Eds.): Innovations in Swarm Intelligence, SCI 248, pp. 141–157.
springerlink.com

One important task in the domain of bioinformatics is genome annotation, particularly the prediction of protein-coding genes. As the blueprint of any organism, genes largely determine the phenotype of the subject through gene expression. Two stages are involved in the process of expressing genes: *transcription* and *translation* where during transcription, the genes in the DNA sequence give rise to *messenger RNA* (mRNA) sequences and in translation, the mRNA is further encoded into a polypeptide chain consisting of multiple amino acids, which eventually folds into the functional gene product — the protein. The annotation of genes involves accurate prediction of several gene signals that help identifying the regions of genomic or mRNA sequences which are used in the process of gene expression.

The translational machinery is mainly comprised of two ribosomal protein subunits and a group of *transfer RNAs* (tRNAs). According to Kozak's famous scanning model hypothesis, the small ribosomal subunit enters the template mRNA from its capped 5' end and advances linearly downstream until it locates the first AUG in a favorable context. This particular location where the translation begins is termed as *translational initiation site* (TIS) or start codon. It is an important type of gene signals in that the accurate recognition of TIS in a messenger RNA sequence leads to the absolute determination of the primary structure of the protein encoded by the gene. The significance of TIS prediction is therefore evident.

Although the prediction of translational start site seems to be straightforward according to the description given above, it is actually more complicated in nature. There are three main reasons for this: (1) *leaky scanning* — the ribosomal subunit that initiates the translation process specifically searches for a 5'-proximal AUG codon with a context that is suitable for translation initiation, which leads to the occurrence of a scenario when an upstream AUG with a poor context is bypassed and a more downstream AUG codon is selected as the start codon; (2) *reinitiation* — this describes a situation when the translation does initiate at the 5'-proximal AUG codon but very soon after the initiation begins an in-frame stop codon is encountered, which results in the production of a short polypeptide chain that may fail to fold into functional gene product; to compensate for this, initiation may be redone when a more downstream AUG codon is chosen as the TIS; (3) *internal ribosome entry site* (IRES) — this happens only on some viral RNA of peculiar structure and it allows the ribosome to initiate in the middle of the mRNA sequence.

Many existing TIS prediction approaches investigate the data from one single perspective or apply some static central fusion mechanism on a fixed set of features. Due to the complicated nature of the genetic data, we believe that it is beneficial to consider multiple biological perspectives in the analysis process. In order to provide diversified problem solving techniques as well as modularization to the system, we have proposed a novel solution that uses a multi-agent system (MAS), which investigates the biological data from multiple biological perspectives, each of which is implemented by an independent problem solver agent with a unique expertise. A generalized layered framework is proposed to facilitate the system to take advantage of the synergy of having multiple agents working together and arriving at a single final prediction. In this chapter, we explore the application of particle swarm optimization and ant colony optimization in the proposed architecture. The reported test results demonstrate the effectiveness of using swarm optimization for TIS prediction.

The rest of the chapter is organized in the following manner. Section 2 presents a brief review to the existing approaches for translational initiation site prediction. In Section 3, a description of the methodology is provided, which includes an overview to the multi-agent system architecture we proposed as well as an introduction to the agents that employ swarm intelligent algorithms. Section 4 reports the experimental results using the proposed approach on three benchmark data sets. In the end, some concluding remarks are given in Section 5.

2 Related Work

The selection of translational initiation sites by the ribosome is believed to be associated with a consensus sequence in most of the cases. Kozak [1] applied a positional weight matrix to examine the conserved context around real start codons and revealed the existence of the following motif in vertebrate messenger RNAs: GCCGCC(A/G)CC<u>AUG</u>G. Salzberg [2] proposed a method that locates the TIS signals by examining the neighborhood of the putative start codons and constructing a conditional probabilistic matrix according to the statistics on pairs of adjacent nucleotide bases.

Pedersen and Nielsen [3] proposed an approach that used artificial *neural network* (NN) to identify the hidden pattern presented by the context of a TIS. This system is the first TIS predictor that successfully applied a machine learning technique for start codon recognition and its rationale has given rise to a variety of computational TIS preditors. Zien *et al.* [4] employed engineered *support vector machines* (SVM) as well as Salzberg kernel to identify the characteristics of a real translational initiation site. Li and Jiang [5] proposed a new sequence similarity measure and experimented with a class of edit kernels for SVMs. Ma *et al.* [6] employed the notion of multiple classifier system which integrates the prediction results from six classifiers each of which is trained with a distinct feature set.

Several approaches have been proposed which apply some feature selection algorithm to identify the most valuable features out of a large raw feature set. One such example is Zeng *et al.*'s work [7] that used a correlation based feature selection scheme on thousands of raw features to arrive at a set of relevant features. They have also experimented with a few different machine learning algorithms and the results are promising. This rationale has also been seen in Liu *et al.*'s approach [8], which used a different set of raw features that mainly reveal the amino acid patterns presented by the sequences. Tzanis and Vlahavas [9] have applied a chi-squared feature selection strategy to extract twenty most relevant elements from a raw feature set. Multiple classifier system has also been used in their system where a weighted majority voting scheme was adopted as the ensemble strategy.

Another category of approaches investigated the applicability of constructing multiple models using different sets of features and combining the outputs from each individual model. Salamov *et al.* [10] applied a *linear discriminant function* (LDF) to analyze six biological features considered relevant to TIS prediction. Hatzigeorgiou [11] used ANNs to investigate both the consensus sequence around AUGs and the coding potential of a putative *open reading frame* (ORF). She then applies a sum rule to yield the final prediction result. Ho and Rajapakse [12] proposed a system with layered structure. The first layer consists of a Markov chain model and a protein encoding model. Its output is then analyzed by three ANNs which are then combined with a majority voting scheme. It was also reported that the scanning model resulted in the improvement of the effectiveness of the approach. Saeys *et al.* [13] explored the possibility of combining three models that applied some of the most primitive TIS prediction tools — position weight matrices, interpolated context models and stop codon frequencies within the putative ORFs and the final prediction is determined by the sum of the outputs from the three participating prediction components.

According to our review to the literature, the most effective start codon prediction systems take advantage of more than one feature, which is in line with the nature of our approach. However, most of these methods use some static fusion strategy such as majority voting or sum function to combine the input from multiple information sources/features (e.g. [10,11,12,13]). Though we agree that the biological translational initiation complex takes several aspects into account when it searches the template mRNA sequence for the actual start codon, we also believe that the selection scheme adopted by the complex is more versatile and adaptive than a static combination strategy like majority voting. To reflect this belief, in our framework, we propose an adaptive algorithm for combining the outputs from multiple predictive agents.

3 Multi-Agent System for Translational Initiation Site Prediction

3.1 General Architecture

Our solution to the problem of translational initiation site prediction is a multi-agent system called *MAS-TIS* that employs multiple types of agents, each of which is designed to accomplish one specific goal. This framework has been originally proposed in our earlier publication [14]. In this section, we begin with a brief presentation of the general architecture and then focus on the introduction to the agents where the swarm intelligent techniques are applied.

According to the framework shown in Fig. 3.1, the MAS-TIS architecture consists of five stages: solution generation, decision making, negotiation, execution and feedback. Four types of agents participate in this entire process — *problem solver agents* (PS), *decision maker agents* (DM), *mediator agent* and *actuator agent*. Problem solver agents are the fundamental agents in the system, which originate a pool of solution candidates. Each of them acts as an independent domain expert which is specialized in solving the problem from a distinct biological perspective and outputs a solution candidate to the solution repository.

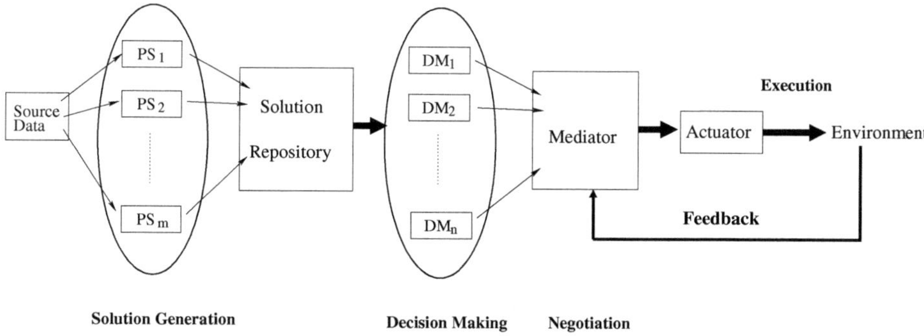

Fig. 1. General Architecture of MAS-TIS

The decision maker agents are introduced in order to take advantage of the synergy provided by multiple domain experts. Each of them employs a unique strategy to manipulate the messages conveyed by the PS agents. Due to the fact that different DM agents may propose conflicting decisions, a mediator agent is needed to arrive at a consistent solution for the system as a whole. This single final decision is then executed by the actuator agent which labels the datum to be *True* or *False* TIS, depending on the instruction of mediator agent. With the purpose of providing adaptive property to the system, the mediator agent is expected to be able to learn from its own experience and evolve over time. To achieve this, during the training phase, a feedback loop is made available which back-propagates the effect of actuating a certain decision on a given datum back to the mediator agent so that it is able to revise its own strategy accordingly if necessary.

3.2 Problem Solver Agent

Each problem solver agent is equipped with a unique set of domain expertise that helps it to provide a solution candidate to facilitate the recognition of TISs. In our current version of the system, three different PS agents have been employed: the *context problem solver* (CPS), which focuses on examining the favorability of the context of a putative start codon; the *downstream problem solver* (DPS), which is trained to examine a putative open reading frame (ORF) in the light of protein secondary structure; and the *codon usage bias problem solver* (CUBPS) which utilizes the codon usage statistics that are observed in most of the genes of the target organism. Details pertaining to these individual agents can be found in [14,15]. In order to provide a general schema that is applicable for solving any functional sequence motif recognition problem, we strive to minimize the application specific elements in our paradigm. To achieve this, we ask each PS agent to produce a *problem solver message* (PSM) that conforms to a standard syntax: (AGENT-NAME: a, $a \in$\{the set of PS agents\}, CLASS: x, $x \in$\{True, False\}).

3.3 Particle Swarm Optimization Based Decision Maker Agent

To offer a separation of PS agents and the subsequent layers of agents in the framework, a layer of decision maker agents are introduced. Some DM agents may solely rely on one PS agent's prediction, thus they output the CLASS prediction offered by the participating PS agent. Since each PS agent is designed to investigate the problem from a local view that matches its own expertise, it may be beneficial to apply some decision maker agents that integrate several PS agents for the purpose of obtaining broader perspective of the problem. For these DM agents, we propose a scheme that applies the weighted sum rule as the ensemble strategy where the optimal weight assignment is acquired by using particle swarm optimization. A quick review to the classical ensemble strategy is in order. Suppose there are n classifiers: $C_1, C_2, \cdots, C_n$. Given a query datum q, each of them predicts q's class membership by outputting a real value. Then the weighted sum of the classifier ensemble C_{Ens} can be computed using Eq. 1.

$$C_{Ens}(q) = \sum_{i=1}^{n} w_i \times C_i(q) \tag{1}$$

where w_i refers to the weight given to $C_i(q)$.

In our system, the PS agents are essentially classifiers and the DM agents that involve using more than one problem solver agent can be considered as classifier ensembles. To obtain the optimal weight vector and provide adaptiveness to the DM agents, we employ the *particle swarm optimization* (PSO) algorithm to evolve the fittest weight assignment scheme. It is also believed that a classifier's prediction capacity on different classes may be unequal. Therefore, more weight can be given to class which the classifier has more confidence predicting.

The studies that inspired the proposal of particle swarm optimization are due to several researchers who were interested in discovering the underlying rules that enabled large number of birds to flock synchronously, often changing direction suddenly, scattering and regrouping, etc.. Notably, Reynolds [16] and Heppner and Grenander [17] had the insight that local processes might underlie the unpredictable group dynamics of bird social behavior. A speculation was proposed which states that individual members of the school can profit from the discoveries and previous experience of all other members of the school during the search for food and shelter.

The algorithm of particle swarm optimization was developed by Russ Eberhart and James Kennedy in 1995 [18]. Since then, the technique has been extensively applied in a spectrum of problems that deal with optimization problems for continuous non-linear functions. The main strategies involved in this algorithm include learning from the success of neighbors and trying to achieve a good balance between exploration and exploitation. Like genetic algorithm, the success of a particular trial solution (represented by a particle in this model) is evaluated by a fitness function. Alg. 1 presents the algorithm of using particle swarm optimization in our system. First of all, a population consisting of many particles is initialized, where a random assignment of velocity and position is given to every particle. Then all of the particles are evaluated by a fitness function. The following two types of positions are recorded to facilitate future sharing of the information: (1) *local best*: for each particle, the position where its highest fitness value is yielded; and (2) *global best*: the position where the maximum fitness value of the entire population is obtained. To simulate the process of learning from the success of one's own and its peers', the velocity of each particle is modified so that its direction is steered towards its local best and the global best. This modification involves the use of weights which are predefined by the users so as to reflect their preference in exploration or exploitation. Subsequently, the positions of all of the particles are recalculated given the new velocities. This process is repeated until convergence is achieved, indicating that a presumptive best solution to the optimization problem is yielded.

A description of using PSO in the problem solver agent ensemble process is in order. During the training phase, to arrive at the optimal weight assignment for decision maker agent DM:$PS_1, PS_2, \cdots, PS_n$, a training data set has to be provided where for each datum D_i, the following vector is available: $(C_{PS_1}(D_i), C_{PS_2}(D_i), \cdots, C_{PS_n}(D_i), C(D_i))$, where $C_{PS_j}(D_i)$ is PS_j's class prediction on D_i and $C(D_i)$ is D_i's actual class, in all cases, class predictions are represented by real numbers (e.g., 1 for *True* and -1 for *False*). First of all, a population of particles are initialized with random positions and velocities, where each particle represents a solution candidate of the optimal weight. Since the goal is to yield the best weight assignment for the multiple classifier ensemble of a binary classification problem, each particle is represented by a vector of $2n$ dimensions. A fitness function is applied which evaluates the suitability of a candidate solution. For any datum D_i, depending on the classes predicted by the participating PS agents, half of the $2n$ dimensions used to represent one particle will be used for calculating the weighted sum of the ensemble. When the numeric result has the same sign as the real number representing the actual class, the fitness of the particle is incremented. This procedure iterates through all of the data in the training set and final fitness value of the particle can be determined. To encourage the particles to learn from the success from the history of their peers' and their own, the positions where each individual had their best fitness are recorded and the global best position is also identified. Subsequently, the velocity of each individual particle is adjusted so that it is steered towards the directions of the global best and the local best. The positions of all of the particles are updated accordingly. If the termination condition is met (e.g., the global best fitness value exceeds a certain threshold or a maximal number of iterations have been reached), the program is terminated and the user is provided with the optimal weight assignment which is the position of the particle that yielded the greatest fitness value. Otherwise, the procedure is repeated.

During the testing phase, for any participating PS agent, depending on its prediction (whether it is positive or negative), a corresponding weight would be used in calculating the overall sum. This total is compared to a threshold value to determine the query datum's class membership. Alg. 1 illustrates the algorithm.

Algorithm 1. Obtain the Fittest Weight Assignment Scheme for DM:$PS_1, PS_2, \cdots, PS_n$ Using PSO

Input: The training set D contains data: $D_1, D_2, \cdots, D_d$. For each datum D_i, the following vector is provided: $(C_{PS_1}(D_i), C_{PS_2}(D_i), \cdots, C_{PS_n}(D_i), C(D_i))$, where $C_{PS_j}(D_i)$ refers to PS_j's class prediction on D_i and $C(D_i)$ corresponds to D_i's actual class, in all cases, class predictions are represented by real numbers (1 for *True* and -1 for *False*).

Output: The fittest weight assignment scheme for the DM agent.

Procedure:

Step 1:

Let each particle correspond to a trial solution to the optimal weight assignment problem and represent it by a $2n$-dimensional vector.
Initialize the population containing N particles by assigning random velocities and positions to them.

$$P = (\boldsymbol{p_1}, \cdots, \boldsymbol{p_N})$$
$$V = (\boldsymbol{v_1}, \cdots, \boldsymbol{v_N})$$

Step 2: Evaluate the fitness of each individual particle p_u by applying the following function: $F(p_u) = \sum_{j=1}^{d} f(\sum_{i=1}^{n} C_{PS_i}(D_j) \times p_{uk}, C(D_j))$, where $f(x,y) = 100$ if $x \times y >= 0$, otherwise, $f(x,y) = 0$; $k = 2i - 1$ if $C_{PS_i} = 1$ and $k = 2i$ if $C_{PS_i} = -1$.

$$F(P) = (F(\boldsymbol{p_1}), \cdots, F(\boldsymbol{p_N}))$$

Step 3: Record the positions where each individual had the best fitness and the maximum out of all.

$$P = (\boldsymbol{p_1}^{\boldsymbol{best}}, \cdots, \boldsymbol{p_N}^{\boldsymbol{best}})$$
$$\boldsymbol{p_g}^{\boldsymbol{best}} = max_{\boldsymbol{p} \in P}(F(\boldsymbol{p}))$$

Step 4: Modify the velocity of each particle according to the global best ($\boldsymbol{p_g}^{\boldsymbol{best}}$) and the local best ($\boldsymbol{p_i}^{\boldsymbol{best}}$) for $i \in [1, N]$

$$\boldsymbol{v_i} = \boldsymbol{v_i} + \phi_1(\boldsymbol{p_i}^{\boldsymbol{best}} - \boldsymbol{p_i}) + \phi_2(\boldsymbol{p_g}^{\boldsymbol{best}})$$

Step 5: Update the particles positions according to the new velocity for $i \in [1, N]$.

$$\boldsymbol{p_i} = \boldsymbol{p_i} + \boldsymbol{v_i}$$

Step 6: Terminate if the convergence condition is achieved.

Step 7: Go to **Step 2**.

3.4 Ant Colony Optimization Based Mediator Agent

Ant colony optimization (ACO) is another example of modeling swarm behavior in computer simulation and yielding impressive gains on a particular set of computational problems. Marco Dorigo originally proposed the algorithm for ACO in his PhD thesis [19], whose aim was to search for an optimal path in a graph. The technique has since been used to solve computer science problems which can be reduced to finding good paths through graphs. The social model that ACO is built upon is an abstraction of the ant colony. In reality, ants wander randomly and upon finding useful resources like food return to their colony while laying down pheromone trails. As a social animal, ants are good at exploiting information provided by their peers. In this case, when an ant encounters a pheromone trail, it usually prefers to follow it rather than picking a different path at random. Like its predecessors, this ant also makes contribution to the trail by laying down pheromone assuming it actually finds the desired resource along the path. As a consequence, the path that is travelled the most tends to have bigger pheromone concentration compared to its less frequently travelled counterparts. On the other hand, pheromone does evaporate, which results in a shorter path to be preferred over a longer path. One of the most well-known applications of ant colony optimization is the *travelling salesman problem* (TSP), whose task is described as follows: given a list of cities and their pairwise distances, find a shortest possible tour that visits each city once and only once. As a classic benchmark combinatorial optimization problem, TSP is shown to be NP-complete. ACO has been used to produce a near-optimal solution to the TSP problem.

In our system, the mediator agent is used to estimate the performance of all of the participating decision maker agents and arrive at a single final decision for the system as a whole. This is accomplished by having the mediator agent select the DM agent presenting the highest performance on predicting a given datum, where performance includes the aspects of both effectiveness and efficiency. We parallel this to the shortest path finding problem which can employ the ant colony optimization algorithm. The algorithm of adapting ACO for the mediator agent is in order. We start with a set of n DM agents and assume that in total m distinct PS agents are used by the DM agent set. The environment is comprised of a single base of origin, m stations representing the PS agents and n ants which are programmed to follow some certain paths. All of the ants are originally located at the original base and travel along some designated path in the hope of finding the resource that is desired. These paths correspond to the DM agents that are considered by the mediator agent and the resource parallels the prediction performance of the DM agent. For instance, assume we have the following decision maker agent: $DM : PS_1, PS_2, PS_3$, which combines the outputs of PS agents 1, 2, and 3. The ant that implements the DM agent then takes the path of traveling from the base to station PS_1, then PS_2 and finally PS_3. Since during the training phase, the DM agent's prediction on any training datum is available, the ant will either find the resource which happens when the prediction is correct or fail to find it when the DM agent's prediction is erroneous.

Algorithm 2. Ant Colony Optimization Based Mediator Agent (Primitive Version)

Input: A set of DM agents: $DM_1, DM_2, \cdots, DM_n$ which employ m distinct PS agents: $PS_1, PS_2, \cdots, PS_m$ and a training set D.

Output: The DM agent that achieves the best balance between effectiveness and efficiency.

Step 1: Locate n ants at the origin base and place m stations at different locations whose distances to the origin are proportional to the corresponding PS agent's computational cost.

Step 2:

for Datum d in D **do**

Let every ant travel its designated path. Upon reaching its destination, obtain the prediction feedback regarding applying the corresponding DM agent on d. If an accurate prediction is made, then the ant returns to the base and leaves a pheromone trail along the way.

Otherwise, the ant goes back to the origin without laying down any chemical.

Simulate the evaporation of pheromone so that the longer of a distance the ant has to travel, the less concentration of pheromone is left. (Note that the chemical content in the pheromone trail is accumulative.)

end for

Step 3: Identify the path with the highest pheromone concentration and output it to the mediator agent as the optimal DM agent to employ.

Regardless of the outcome of the pursuit of a certain path, the ant returns to the base after it travels the path in its entirety, and if after following the path, it is provided with the desired resource, the ant then leaves a pheromone trail on its returning trip. The next datum in the training collection will then be used by this set of ants which travel the designated path once again. However this time the outcome may be different since the datum is different from the one used in the previous iteration. To find the most effective yet efficient DM agent, a shortest path needs to be identified. This is achieved by placing the PS agents strategically so that the more computational costly agents are mapped to the stations that are further away from the base and the pheromone's evaporative property is integrated into the implementation so that after all of the training data are exhausted, the path that has the highest concentration of pheromone is considered as the optimal solution therefore its corresponding DM agent is identified. Alg. 2 illustrates the algorithm.

As mentioned in Section 2, in our system, the individual characteristics of different data are recognized. Therefore the DM agent that corresponds to a shortest path yielded by applying all of the training data does not necessarily provide the best solution to all of the data in the testing set. Our solution to this is to apply a minor extension to the algorithm described above. Given the participating PS agents, we construct a database's key space consisting of m dimensions (where m is the number of problem solver agents that are used in the DM agent set). Since the TIS prediction is a binary prediction problem, there exist 2^m distinct keys ranging from $(True, True, \cdots, True)$ to $(False, False, \cdots, False)$. In this way, the data are divided into 2^m groups depending on their key value. Alg. 2 is then applied to each group so as to locate the best DM agent for the group. During the testing phase, a query datum goes through the solution generation process and its corresponding key can be obtained based upon the predictions of the PS agents. The most appropriate DM agent for this key entry can be easily retrieved from the database, which is then applied by the mediator agent to make the final prediction on the membership of the query datum.

4 Experiments

4.1 Data Sets

In order to evaluate the effectiveness of the proposed approach, we conducted testing using three benchmark data sets — *vertebrates*, *Arabidopsis thaliana* and *TIS+50*. The sequences from the first two data collections were extracted from GenBank, release 95. All of the sequences have undergone preprocessing so that possible introns were removed and only the sequences that contain at least 10bp upstream of the TIS and at least 150bp downstream of the TIS were selected. The vertebrates group consists of sequences from *Bos taurus* (cow), *Gallus gallus* (chicken), *Homo sapiens* (human), *Mus musculus* (mouse), *Oryctolagus cuniculus* (rabbit), *Rattus norvegicus* (rat), *Sus scrofa* (pig), and *Xenopus laevis* (African clawed frog). The second data set contains sequences from *Arabidopsis thaliana* (thale cress, a dicot plant), which presents large deviation from vertebrates. *TIS+50* contains 50 human *expressed sequence tags* (EST) sequences with complete ORFs. Table 1 summarizes the characteristics of the data sets.

Table 1. Data Sets

Name	Authors	# of Positives	# of Negatives
vert.	[3]	3312	10191
Arab.	[3]	523	1525
TIS+50	[20]	50	439

There are several reasons why we selected these three data sets as the testing sets in our experiments. Firstly, all of these collections have been used to test the effectiveness of more than one existing algorithm, especially the vertebrates data set, which has been cited in most of the related work. Secondly, the vertebrates and Arabidopsis collections only include conceptually-spliced mRNAs whereas TIS+50 contains EST sequences that may contain errors resulting in frame shifts, and represent different parts of their parent cDNA. The difference between the two types of sequences provides some diversity to the testing process. The availability factor also plays a part in making our decision — all of the data sets that are used in our paper are easily downloadable from the Internet[1].

4.2 Evaluation Criteria

We have applied three-fold cross validation in all of the experiments that are conducted. The following four are used in our study — *sensitivity* (Sen), *specificity* (Spe), *adjusted accuracy* (AA) and *overall accuracy* (OA). Table 2 shows a contingency table. Given the context of TIS prediction, sensitivity refers to the ratio of the number of correctly identified TISs (labeled by *True*) over the total number of TISs, a high sensitivity implies that a big percentage of TISs have been correctly identified by the predictor; specificity refers to the ratio of correctly identified pseudo-TISs (labeled by *False*) over the total number of pseudo-TISs, a high specificity indicates that a substantial percentage of pseudo-TISs have been accurately recognized by the predictor. It is easy to notice that both of these metrics have their own evaluation bias, one focuses on the prediction performance of the true data whereas the other focuses on that of the false data. Adjusted accuracy, which is the average of sensitivity and specificity, serves as a comprehensive evaluation measure that tries to be as neutral as possible. It is also a better choice than the metric of overall accuracy especially for dealing with skewed data sets, e.g., the TIS data collections. In our experiments, adjusted accuracy will be considered as the definitive metric and all of the comparisons will be conducted based on it unless suggested otherwise.

Table 2. Contingency Table

	Classified as True	Classified as False
Actual True	TP	FN
Actual False	FP	TN

The formulas that are used to compute each of the four metrics are listed as follows:

$$Sen = \frac{TP}{TP + FN}$$
$$Spe = \frac{TN}{TN + FP}$$
$$AA = \frac{Sen + Spe}{2}$$
$$OA = \frac{TP + TN}{TP + FP + TN + FN}$$

[1] For NetStart: http://www.cbs.dtu.dk/services/NetStart/.
For TIS+50: http://www.biomedcentral.com/content/supplementary/1471-2105-5-14-S1.txt.

4.3 Particle Swarm Optimization Based Decision Maker Agent

As one of the novel components proposed in this paper, we are interested in evaluating the effectiveness of the PSO based decision maker agents by itself. An open source PSO library called JSwarm-PSO [21] is used in our implementation. To facilitate the performance analysis of the decision maker agent that uses the particle swarm optimization as the optimal weight acquisition scheme, we simplify the architecture by eliminating the mediator agent and by only using one decision maker agent in the system. In Table 3, the rows entitled PSO (CPS-DPS) and PSO (CPS-DPS-CUBPS) report the results using PSO to obtain the optimal weights for DM:CPS-DPS and DM:CPS-DPS-CUBPS respectively. To investigate the merit of the PSO based scheme, a comparative study is necessary. We have included severals rows entitled Max (CPS-DPS) which report the statistics related to the decision maker agent DM:CPS-DPS that uses a max rule to combine the outputs of the two PS agents. In some variant of the PS agent, the confidence of its prediction is provided. The max rule simply outputs the CLASS prediction offered by the PS agent that has the highest confidence value. Since CUBPS has not implemented such a confidence evaluation scheme, the max rule cannot be applied to any DM agent that uses CUBPS. This explains why Max (CPS-DPS-CUBPS) is missing from the table. From the data shown in Table 3, in both *vert.* and *Arab.*, the PSO based DM agent significantly outperforms the max rule counterpart when only CPS and DPS are used by the DM agent under investigation whereas in *TIS+50* the former performs slightly worse than the latter. The applicability and effectiveness of using particle swarm optimization algorithm for our decision maker agent are then demonstrated.

4.4 Ant Colony Optimization Based Mediator Agent

To estimate the effectiveness of the mediator agent that employs ant colony optimization, we conduct experiments using three PS agents: CPS, DPS, and CUBPS and the corresponding DM agent set together with the ACO based mediator agent. Table 4 reports the complete results.

Table 3. Evaluation on Particle Swarm Optimization Based Decision Maker Agent

Data Set	DM Scheme	Sen	Spe	AA	OA
vert.	Max (CPS-DPS)	55.28%	97.79%	76.54%	87.28%
	PSO (CPS-DPS)	99.12%	64.42%	81.77%	73.03%
	PSO (CPS-DPS-CUBPS)	93.00%	97.40%	95.20%	96.32%
Arab.	Max (CPS-DPS)	90.06%	95.90%	92.98%	94.36%
	PSO (CPS-DPS)	97.32%	95.90%	96.61%	96.23%
	PSO (CPS-DPS-CUBPS)	97.51%	96.39%	96.95%	96.67%
TIS+50	Max (CPS-DPS)	37.75%	98.72%	68.23%	91.93%
	PSO (CPS-DPS)	37.87%	96.93%	67.40%	90.61%
	PSO (CPS-DPS-CUBPS)	72.30%	95.99%	84.15%	93.50%

Table 4. Complete Testing of MAS-TIS using ACO Based Mediator Agent

Data Set	Sen	Spe	AA	OA
vert.	96.23%	95.58%	95.90%	95.73%
Arab.	91.40%	96.41%	93.91%	95.06%
TIS+50	84.07%	96.95%	90.51%	95.63%

To put the data into perspective, we completed a comparative study that involves the following seven existing state-of-the-art methods for TIS prediction: Pedersen and Nielsen's NetStart system [3][2], GENSCAN system [22][3], Saeys *et al.*'s StartScan system [13][4], Salzberg's positional conditional probability approach [2], Zien *et al.*'s engineered SVM method [4], Zeng *et al.*'s system [7], Liu *et al.*'s method [8] and Salamov *et al.*'s LDF scheme.

The first three systems exist in the form of online servers, therefore complete results using all of the three data collections can be obtained. In particular, GENSCAN is considered as one of the most successful coding sequence recognition approaches merely relying on properties intrinsic to nucleotide sequences. Most of the remaining existing approaches report the performance of their systems using *vert.* collection alone except for Salamov *et al.*'s LDF scheme, which is tested by Nadershahi *et al.* [20] and the collection used is *TIS+50*. With the exception of GENSCAN, descriptions to all of the aforementioned related work can be found in Section 2.

Since for some of the existing systems, only one measure is used which is the overall accuracy, we consider both adjusted accuracy and overall accuracy in the comparison. Table 5 reports all of the available results regarding the approaches of interest, where the highest accuracies are shown in bold font. From the data we can observe that throughout all of the methods under investigation, the best performance is always yielded by our MAS-TIS system that uses ACO based mediator agent, which significantly outperforms the existing approaches. The effectiveness and robustness of the proposed approach have been well demonstrated. To shed some light on how MAS-TIS achieves its outstanding performance, a more in-depth comparative analysis is in order.

Like most other statistical based approaches, Salzberg's method suffers from high false positive rate. Our MAS-TIS approach integrates a variety of algorithms that diversify the solution set. Although Pedersen and Nielsen were the pioneers of applying machine learning algorithm to solve TIS prediction problems, by considering only the context information around the AUG codons, their strategy has oversimplified the underlying biological process. We believe that the region that is upstream to a putative TIS should be considered differently from the one that is downstream to an AUG. As well, according to Kozak's ribosomal scanning model, the relative position of an AUG in an mRNA sequence also plays a critical

[2] http://www.cbs.dtu.dk/services/NetStart/
[3] http://genes.mit.edu/GENSCAN.html
[4] http://bioinformatics.psb.ugent.be/webtools/startscan/

Table 5. Comparative Study

Data Set	Method	AA	OA
vert.	MAS-TIS	**95.90%**	**95.73%**
	NetStart	85.02%	86.44%
	GENSCAN	45.24%	68.17%
	StartScan	69.06%	63.57%
	Salzberg	70.9%	86.2%
	Zien *et al.*	77.2%	88.6%
	Zeng *et al.*	92.4%	94.4%
	Liu *et al.*	88.34%	92.45%
Arab.	MAS-TIS	**93.91%**	**95.06%**
	NetStart	93.06%	90.97%
	GENSCAN	44.94%	66.65%
	StartScan	50.56%	63.33%
TIS+50	MAS-TIS	**90.51%**	**95.63%**
	NetStart	78.97%	71.78%
	GENSCAN	81.20%	94.89%
	StartScan	70.03%	56.18%
	Salamov *et al.*	—	90%

role in determining the fitness of a TIS candidate. Using a similar strategy as Pedersen and Nielsen's method, Zien *et al.*'s SVM prediction model also presents similar limitations. In MAS-TIS, we have incorporated a more versatile set of biological aspects, each investigated by an independent problem solver agent.

Some other existing approaches share this perspective of combining multiple models to solve the problem. For instance, Salamov *et al.* uses a linear discriminant function to combine six different features of the data, Saeys *et al.*'s StartScan system applies a plain sum rule to combine the output from three independent classifiers, Zeng *et al.* and Liu *et al.* both use some machine learning algorithm that takes all of the relevant features as the feature set. Regardless of the differences in the feature set, these approaches share one common problem, i.e., the decision fusion strategy is static. In MAS-TIS, during the decision making process, we incorporate a particle swarm optimization based decision maker scheme that takes advantage of the adaptive property presented by the swarm intelligence scheme; in the negotiation process, each participating DM agent is evaluated based upon its potential profitability which is modeled by an ant in an ant colony environment and depending on the outputs of the problem solver agents, different DM agents will be considered to be optimal ensemble strategy. Such a configuration provides a good flexibility and adaptability to the system as a whole. It is also worth mentioning that despite GENSCAN's success in common gene prediction, it is not solely designed to predict TIS. Its corresponding TIS prediction results are yielded by applying an indirect procedure using the output from the program. This may explain its relatively unsatisfactory performance on some of the testing sets reported in this paper.

5 Conclusion

In this chapter, we have investigated an important topic in bioinformatics — the prediction of translational initiation sites in genomic, mRNA and cDNA sequences. A novel approach that uses a multi-agent architecture is introduced. Two famous swarm intelligent algorithms including particle swarm optimization and ant colony optimization have been incorporated into the framework where PSO is used to obtain the optimal weight assignment scheme for the decision maker agents which use weighted sum rule for classifier ensemble and ACO is employed to facilitate the mediator agent to arrive at a reasonable estimation of the performance of all of the DM agents that are involved. Both swarm intelligence approaches are well-established algorithms that present adaptive properties. The application of these methods results in the outstanding performance of our multi-agent system, which can be seen from the results of the experiments conducted on three benchmark data sets.

References

1. Kozak, M.: An analysis of 5'-noncoding sequences from 699 vertebrate messenger RNAs. Nucleic Acids Research 15(20), 8125–8148 (1987)
2. Salzberg, S.: A method for identifying splice sites and translational initiation sites in eukaryotic mRNA. Computer Applications in the Biosciences 13, 365–376 (1997)
3. Pedersen, A., Nielsen, H.: Neural network prediction of translation initiation sites in eukaryotes: Perspectives for EST and genome analysis. In: Proceedings of the Fifth International Conference on Intelligent Systems for Molecular Biology, pp. 226–233 (1997)
4. Zien, A., Ratsch, G., Mika, S., Scholkopf, B., Lemmen, C., Smola, A., Lengauer, T., Muller, K.: Engineering support vector machine kernels that recognize translation initiation sites. Bioinformatics 16(9), 799–807 (2000)
5. Li, H., Jiang, T.: A class of edit kernels for SVMs to predict translation initiation sites in eukaryotic mRNAs. Journal of Computational Biology 12(6), 702–718 (2005)
6. Ma, C., Zhou, D., Zhou, Y.: Feature mining and integration for improving the prediction accuracy of translation initiation sites in eukaryotic mRNAs. In: Proceedings of the Fifth International Conference on Grid and Cooperative Computing Workshop, pp. 349–356 (2006)
7. Zeng, F., Yap, R., Wong, L.: Using feature generation and feature selection for accurate prediction of translation initiation sites. Genome Informatics 13, 192–200 (2002)
8. Liu, H., Han, H., Li, J., Wong, L.: Using amino acid patterns to accurately predict translation initiation sites. In Silico. Biology 4(3), 255–269 (2004)
9. Tzanis, G., Vlahavas, I.: Prediction of translation initiation sites using classifier selection. In: Antoniou, G., Potamias, G., Spyropoulos, C., Plexousakis, D. (eds.) SETN 2006. LNCS (LNAI), vol. 3955, pp. 367–377. Springer, Heidelberg (2006)
10. Salamov, A., Nishikawa, T., Swindells, M.: Assessing protein coding region integrity in cDNA sequencing projects. Bioinformatics 14(5), 384–390 (1998)
11. Hatzigeorgiou, A.: Translation initiation start prediction in human cDNAs with high accuracy. Bioinformatics 18(2), 343–350 (2002)

12. Ho, L., Rajapakse, J.: High sensitivity technique for translation initiation site prediction. In: 2004 IEEE Symposium on Computational Intelligence in Bioinformatics and Computational Biology, pp. 153–159 (2004)
13. Saeys, Y., Abeel, T., Degroeve, S., de Peer, Y.: Translation initiation site prediction on a genomic scale: beauty in simplicity. Bioinformatics 23 ISMB/ECCB 2007, i418–i423 (2007)
14. Zeng, J., Alhajj, R.: Predicting translation initiation sites using a multi-agent architecture empowered with reinforcement learning. In: 2008 IEEE Symposium on Computational Intelligence in Bioinformatics and Computational Biology, pp. 241–248 (2008)
15. Zeng, J., Alhajj, R., Demetrick, D.: The effectiveness of applying codon usage bias for translational initiation sites prediction. In: 2008 IEEE International Conference on Bioinformatics and Biomedicine, pp. 121–126 (2008)
16. Reynolds, C.: Flocks, herds and schools: a distributed behavioral model. Computer Graphics 21(4), 25–34 (1987)
17. Heppner, F., Grenander, U.: The Ubiquity of Chaos. In: A stochastic nonlinear model for coordinated bird flocks. AAAS Publications, Washington (1990)
18. Kennedy, J., Eberhart, R.: Particle swarm optimization. In: Proceedings of the IEEE International Conference on Neural Networks, Piscataway, NJ, pp. 1942–1948 (1995)
19. Dorigo, M.: Optimization, Learning and Natural Algorithms. PhD thesis, Politecnico di Milano (1992)
20. Nadershahi, A., Fahrenkrug, S., Ellis, L.: Comparison of computational methods for identifying translation initiation sites in EST data. BMC Bioinformatics 5(14) (2004)
21. `http://jswarm-pso.sourceforge.net/`
22. Burge, C., Karlin, S.: Prediction of complete gene structures in human genomic DNA. Journal of Molecular Biology 268(1), 78–94 (1997)

9

Particle Swarm Optimization for Optimal Operational Planning of Energy Plants

Yoshikazu Fukuyama, Hideyuki Nishida, and Yuji Todaka

Fuji Electric Systems Co., Ltd., No.1, Fuji-machi, Hino-city, Tokyo 191-8502, Japan
fukuyama-yoshikazu@fesys.co.jp

Abstract. In this chapter, three PSO based methods: Original PSO, Evolutionary PSO, and Adaptive PSO are compared for optimal operational planning problems of energy plants, which are formulated as Mixed-Intger Nonlinear Problems (MINLPs). The three methods are compared using typical energy plant operational planning problems. We have been developed an optimal operational planning and control system of energy plants using PSO (called FeTOP). FeTOP has been actually introduced and operated at three factories of one of the automobile company in Japan and realized 10% energy reduction compared with operators' operation.

1 Introduction

Recently, cogeneration systems (CGS) have been installed in energy plants of various factories and buildings. CGS is usually connected to various facilities such as refrigerators, reservoirs, and cooling towers. It produces various energies including electric loads, air-conditioning loads, and steam loads (figure 1). Since daily load patterns of the loads are different, daily optimal operational planning for an energy plant is a very important task for saving operational costs and reducing environmental loads. In order to generate optimal operational planning for an energy plant, various loads should be forecasted, and startup and shutdown status and input values for the facilities at each control interval should be determined using facility models (figure 2). Therefore, the optimal operational planning problem can be formulated as a mixed-integer linear problem (MILP) and mathematical programming techniques such as branch-and-bound methods, decomposition methods, and dynamic programming have been applied conventionally [1-3]. However, the facilities may have nonlinear input-output characteristics practically, and operational rules, which cannot be expressed as mathematical forms, should be considered in actual operation. For example, steam turbines usually have plural boilers and the characteristics of the turbines cannot be expressed with only one equation and the characteristics should be expressed with combination of equations with various conditions. Therefore, when the models are constructed, we should utilize concept of data mining. In addition, in the problem, various objectives should be considered such as reduction of operational costs and environmental loads. Consequently, the problem cannot be solved by the conventional methods and the method for solving the multi-objective MINLP problem with complex models has been eagerly awaited.

C.P. Lim et al. (Eds.): Innovations in Swarm Intelligence, SCI 248, pp. 159–173.
springerlink.com

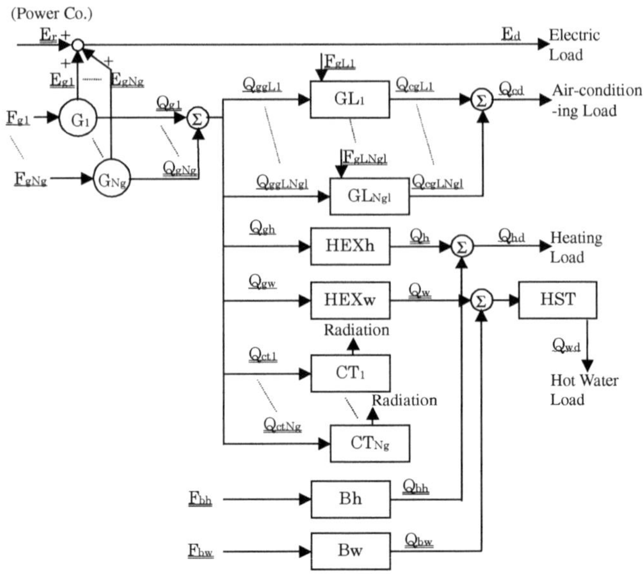

Ng : Number of generator (Gen), Ngl : Number of genelink, G : Gen.,
GL : Genelink, HEXh : Heat exchanger (HEX) for heat load,
HEXw : HEX for hot water load, HST : Hot water storage tank
CT : Cooling tower, Bh : Boiler for heating, Bw : Boiler for water supply,
E_r : Receiving/sending electric energy,
F_g : CGS fuel, E_g : CGS electric power output, E_d : Electric load,
Q_g : CGS heat output, Q_{ggL} : GL input heat energy, F_{gL} : GL fuel,
QcgL : GL output heat energy, Q_{cd} : Air-conditioning load,
Q_{gh} : HEXh input heat energy, Q_h : HEXh output heat energy,
Q_{hd} : Heat load, Q_{gw} : HEXw input heat energy, Q_w : HEXw output heat energy,
Q_{wd} : Hot water load, F_{bh} : Bh fuel, Q_{bh} : Bh output heat energy, F_{bw} : Bw fuel,
Q_{bw} : Bw output heat energy, Q_{ct} : Radiation value.

Fig. 1. A typical CGS system.

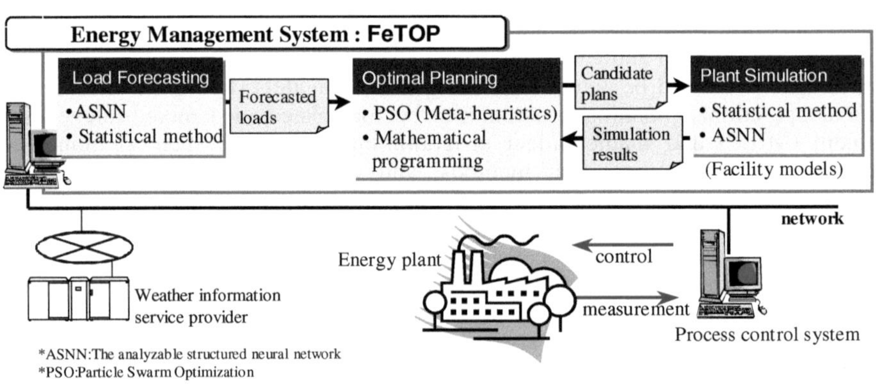

Fig. 2. A basic concept of the optimal operational planning for energy plants.

Particle Swarm Optimization (PSO) is one of the evolutionary computation techniques [4]. PSO is suitable for the optimal operational planning for energy plants because it can handle multi-objectives, on-site operation rules, constraints, and independent complex facility models easily. The original PSO was developed by Eberhart and Kennedy [4]. Recently, various modified methods have been developed and applied to various problems [5-12]. We have been developed an optimal operational planning and control system of energy plants using PSO (called FeTOP) [13][14]. FeTOP has been actually introduced and operated at three factories of one of the automobile companies in Japan and realized 10% energy reduction [15]. Forecasting various loads is out of scope in this paper. However, we have already developed the analyzable structured neural network (ASNN) and other forecasting methods. The accurate load forecasting can be realized for various loads [16]. When we construct forecasting models, data mining methods should be used so that the difference of models such as weekdays and weekends can be treated.

In this chapter, three PSO based methods: Original PSO, Evolutionary PSO, and Adaptive PSO are compared for optimal operational planning problems of energy plants. The target problem is formulated as MINLPs. The three methods are compared using typical energy plant operational planning problems.

2 Problem Formulation

2.1 State Variables

State variables are electrical power output values of generator, heat energy output values of genelink and heat exchanger, and heat energy input values of genelink per hour (24 points a day). The detailed variables and related boundary constraints can be listed as follows:

(1) Generator

The state variables of generators are as follows:

P_{gni} : Electrical power output (24 points a day)

$\delta_{gi} \in \{0,1\}$: Startup/shutdown status

where,

$P_{gn\min} \le P_{gni} \le P_{gn\max}$ (i=0,..,23, n=1,..,Ng) ,

N_g : Number of generator,

P_{gnmax} : Maximum output,

P_{gnmin} : Minimum output.

(2) Genelink

Genelink is a kind of absorption refrigerators, which can decrease adding fuels by inputting wasted heat energy of generator. The state variables of genelink are as follows:

Q_{ggLni} : Heat input values (24 points a day)

Q_{cgLni} : Output heat values (24 points a day)

$\delta_{gli} \in \{0,1\}$: Startup/shutdown status

where,

$0 \leq Q_{ggLni} \leq Q_{ggLni\max}$ $(i=0,..,23,\ n=1,..,N_{gl}),$

$0 \leq Q_{cgLni} \leq \min(Q_{cdi}, Q_{rcgLn})$ $(i=0,..,23,\ n=1,..,N_{gl}),$

N_{gl} : Number of genelink,

$Q_{ggLni\max}$: Maximum heat input values determined by output heat values,

Q_{cdi} : Air-conditioning load,

Q_{rcgLn}: Rated air-conditioning output.

(3) Heat Exchanger

The state variables of heat exchanger for heating/hot water supply are as follows:

(a) Heat Exchanger for Heating

Q_{ghni} : Heat energy input values (24 points a day)

$\delta_{hexhi} \in \{0,1\}$: Startup/shutdown status

where,

$0 \leq A_{hexhn} Q_{ghni} \leq Q_{hdi}$ $(i=0,..,23,\ n=1,..,N_{hexh}),$

N_{hexh} : Number of heat exchanger for heating,

Q_{hdi} : Heat load,

A_{hexhn} : Coefficients of facility characteristics.

(b) Heat Exchanger for Hot Water Supply

Q_{gwi} : Heat energy input values (24 points a day)

$\delta_{hexwi} \in \{0,1\}$: Startup/shutdown status

where,

$0 \leq A_{hexwn} Q_{gwni} \leq Q_{wdi}$ $(i=0,..,23,\ n=1,..,N_{hexw}),$

N_{hexw} : Number of heat exchanger for hot water supply,

Q_{wdi} : Hot water supply load,

A_{hexwn} : Coefficients of facility characteristics.

Outputs of each facility for 24 points of the day should be determined. Moreover, two or three variables are required for one facility (startup and shutdown status (binary or discrete variables), and output or output/input values (continuous variables)). Therefore, one state variable for one facility is composed of vectors with 48 (24 points x 2 variables) or 72 (24 points x 3 variables) elements. Therefore, for example, handling two generators, two genelinks, and two heat exchangers require 336 variables.

2.2 Objective Function

The objective function is to minimize the operational costs and environmental loads of the day.

$$\min\ w_1(CE + Cg + Cw) + w_2 EL \tag{1}$$

where,

CE : Total electricity charges of a day,

Cg : Total fuel charges of a day,

Cw : Total water charges of a day,

EL: Environmental loads of a day,

w_i : weighting factors.

2.3 Constraints

The following constraints are considered in the target problem:

(1) Demand and supply balance: Summation of energies supplied by facilities such as electrical power, air-conditioning energy, and heat energy should be equal to each corresponding load.

(a) Electric Energy Balance

Summation of purchase or sale electric energies and electric power generation values by CGS should be equal to electric loads:

$$E_{ri} + \sum_{n}^{Ng} E_{gni} = E_{di} \quad (i=0,..,23) \tag{2}$$

where, E_{ri} : Purchase or sale electric energies,
E_{gni}: Electric power generation values,
E_{di} : Electric loads.

(b) Air-conditioning Energy Balance

Summation of air-conditioning energies should be equal to air-conditioning loads.

$$\sum_{n=1}^{N_{gl}} Q_{cgLni} = Q_{cdi} \quad (i=0,..,23) \tag{3}$$

(c) Heat Energy Balance

Summation of heat energy inputs and heat energies produced by boilers should be equal to heat loads.

$$Q_{hi} + \sum_{n=1}^{N_{bh}} Q_{bhni} = Q_{hdi} \quad (i=0,..,23) \tag{4}$$

$$Q_{hi} = \sum_{n=1}^{N_{hexh}} A_{hexhn} Q_{ghni} \quad (i=0,..,23) \tag{5}$$

where, N_{bh} : Number of boiler for heating,
Q_{bhni} : Output of boiler for heating.

(d) Hot Water Supply Balance

Summation of hot water inputs and hot waters produced by boilers should be equal to hot water loads.

$$Q_{wi} + \sum_{n=1}^{N_{bw}} Q_{bwni} = Q_{wdi} \quad (i=0,..,23) \tag{6}$$

$$Q_{wi} = \sum_{n=1}^{N_{hexw}} A_{hexwn} Q_{gwni} \quad (i=0,..,23) \tag{7}$$

where, N_{bw} : Number of boiler for hot water supply.

(e) Heat Balance

Summation of the heat energy consumptions at genelinks, for heat and hot water loads, and radiation values at cooling towers should be equal to the (wasted) heat energy produced by CGSs.

$$\sum_{n=1}^{N_g} Q_{gni} = \sum_{n=1}^{N_{gl}} Q_{ggLni} + \sum_{n=1}^{N_{hexh}} Q_{ghni} + \sum_{n=1}^{N_{hexw}} Q_{gwni} + \sum_{n=1}^{N_g} Q_{ctni} \tag{8}$$

$Q_{gni}^{k} = f\left(F_{gni}\right)(i=0,..,23)$

where, Q_{ctni} : Radiation value at cooling tower,

Q_{gni} : Heat output of generator.

(2) Facility constraints: Various facility constraints including the boundary constraints with state variables should be considered. Input-output characteristics of facilities should be also considered as facility constraints. For example, the characteristic of genelink is nonlinear practically and the nonlinear characteristic should be considered in the problem.

(3) Operational rules: Various operational rules should be considered. The followings are examples of the rules:

- If the facility is startup, then the facility should not be shutdowned for a certain period. (Minimum up time)
- If the facility is shutdowned, then the facility should not be startup for a certain period. (Minimum down time)

Facility models are constructed using the facility constraints and the operational rules. The models are independent and all states of the energy plant are calculated when all of facility states are input from PSO. Then, the operational costs and the environmental loads for the days can be calculated. When we construct facility models using actual operation data, we have to construct plural models even for one facility using data mining concepts because we have to utilize various operating points for facilities so that supply and demand balance of various energies should be maintained. Namely, actual operation data can be divided into several groups and facility models are constructed for each group using data mining concepts.

3 Particle Swarm Optimization

3.1 Original PSO [4][5]

The original PSO is a population based stochastic optimization technique developed by Kennedy and Eberhart[4][5]. The current searching points are modified using the following state equations.

$$v_i^{k+1} = wv_i^k + c_1 r_1 \times (pbest_i - s_i^k) + c_2 r_2 \times (gbest - s_i^k) \tag{9}$$

$$s_i^{k+1} = s_i^k + v_i^{k+1} \tag{10}$$

where,

v_k^i : Velocity of particle i at iteration k,

w : Weighting function,

c_i : Weighting coefficients,

r_i : Random number between 0 and 1,

s_k^i : Current position of particle i at iteration k, $pbest_i$: pbest of particle i, $gbest$: gbest of the group.

The original PSO algorithm can be expressed as follows

1) *State variables (searching point)* : State variables (states and their velocities) can be expressed as vectors of continuous numbers. PSO utilizes multiple searching points as agents for search procedures.
2) *Generation of initial searching points* (Step.1) : Initial conditions of searching points in the solution space are usually generated randomly within their allowable ranges.
3) *Evaluation of searching points* (Step.2) : The current searching points are evaluated by the objective function of the target problem. *Pbests* (the best evaluated value so far of each agent) and *gbest* (the best of pbest) can be modified by comparing the evaluation values of the current searching points, and current pbests and gbest.
4) *Modification of searching points* (Step.3) : The current searching points are modified using the state equations of PSO.
5) *Stop criterion* (Step.4) : The search procedure can be stopped when the current iteration number reaches the predetermined maximum iteration number. For example, the last gbest can be output as a solution.

3.2 Evolutionary PSO (EPSO)[9][10]

The idea behind EPSO is to grant a PSO scheme with an explicit selection procedure and with self-adapting properties for its parameters. At a given iteration, consider a set of solutions or alternatives, particles are modified and reproduced. The general scheme of EPSO is the following:

1)REPLICATION - each particle is replicated R times,
2)MUTATION - each particle has its mutated weights,
3)REPRODUCTION - each mutated particle generates an offspring according to the particle movement rule,
4)EVALUATION - each offspring has its fitness evaluated,
5)SELECTION - by stochastic tournament the best particles survive to form a new generation.

The velocity of the state equations for EPSO is the following:

$$v_i^{k+1} = w_{i0}^* v_i^k + w_{i1}^* \left(pbest_i - s_i^k \right) + w_{i2}^* \left(gbest^* - s_i^k \right) \tag{11}$$

So far, this seems like PSO; the movement rule keeps its terms of inertia, memory and cooperation. However, the weights undergo mutation

$$w_{ik}^* = w_{ik} + \tau N(0,1) \tag{12}$$

Where, *N(0,1)* is a random variable with Gaussian distribution, 0 mean and variance 1. The τ, are learning parameters.

3.3 Adaptive PSO (APSO)[11][12]

The adaptive PSO is based on the results of the analysis and the simulations on the basis of the stability analysis in discrete dynamical systems. The new parameters (*p*) are set to each particle. The weighting coefficients are calculated as follows.

If the particle becomes pbest;

$$c_2 = \frac{2}{p}, \quad c_1 = any\ . \tag{13}$$

If the particle is not pbest;

$$c_2 = \frac{1}{p}, \quad c_1 = c_2 \cdot \frac{|gbest - s|}{|pbest - s|}\ . \tag{14}$$

The search trajectory of PSO can be controlled by the parameter (p). Concretely, when the value is enlarged more than 0.5, the particle may move close to the position of gbest.

The adaptive procedure can be expressed as follows:

1) If the particle becomes gbest, the weighting coefficient (w) is set to 1. Otherwise, w is set to 0.
2) When the particle goes out the feasible region, the parameter (p) is set to 0.5 or more so that the particle may convergence.
3) When the objective function value of pbest of the particle is improved for a certain period, the parameter (p) is reduced more than 0.5 so that the particle may divergence.

3.4 Expansion of PSO for Optimal Operational Planning

In order to reduce the number of state variables, the following simple expansion of PSO is utilized. Namely, all of state variables can be expressed as continuous variables. If the input value for a facility is under the minimum input value, then the facility is recognized as shutdown. Otherwise, the facility is recognized as startup and the value is recognized as the input of the facility. The reduction method can reduce the number of state variables to half, and drastic improvement of PSO search procedures can be expected.

4 Optimal Operational Planning for Energy Plants Using PSO

All of state variables have 24 elements and one state in the solution space can be expressed as an array with number of all facilities multiplied by 24 elements. A flow chart is shown in figure 3. The whole algorithm can be expressed as follows:

Step.1: *Generation of initial searching points (states)*: States and velocities of all facilities are randomly generated. The upper and lower bounds of facilities are considered.

Step.2: *Evaluation of searching points*: The current states are input to facility models and the total operational costs are calculated as the objective function value. The *pbests* and *gbest* are updated based on the value.

Step.3: *Modification of searching points*: The current searching points (facility states) are modified using the state equations. The upper and lower bounds of facilities are considered when the current states are modified.

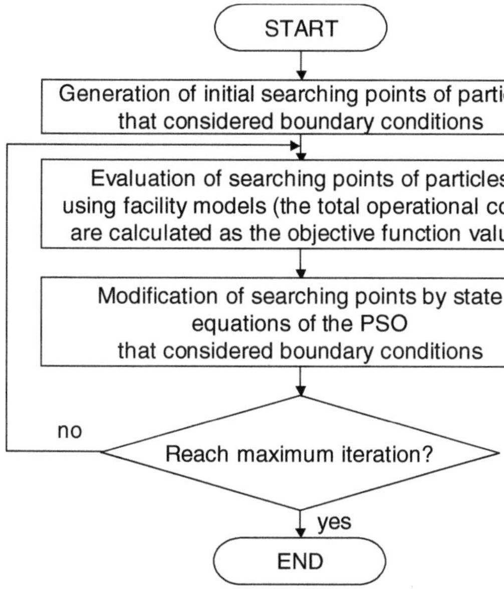

Fig. 3. A flow chart of optimal operational planning for energy plants using PSO.

Step.4: *Stop criterion*: The search procedure can be stopped when the current iteration number reaches the predetermined maximum iteration number. Otherwise, go to step. 2. The last *gbest* is output as a solution.

5 Numerical Examples

5.1 A typical CGS system

(1) Simulation Conditions

An office load model with 100000 [m2] total floor spaces is utilized in the simulation. Two CGS generators (750 [kW]/unit) and two genelinks (4700 [kW]/unit) are assumed to be installed. At most, two genelinks can be startup in summer season, one genelink in winter season, and one genelink in intermediate season. The efficient rate of the heat exchanger is 0.975 and the rate of the boiler is 0.825. The rated capacity of the cooling tower is 1050 [kW]/unit. The cooling tower is installed for each CGS. The forecasted loads for three represented days are shown in figure 4 - figure 6. Only daily costs are considered in the simulation for the sake of simplicity. Number of particles is set to 200. The iteration number is set to 100. Twenty trials are compared. The numbers may be able to be optimized and the further investigation should be performed.

(2) Simulation Results

Table 1 shows comparison of costs by three PSO based methods and the conventional rule based planning method. According to the results, the total operational cost is

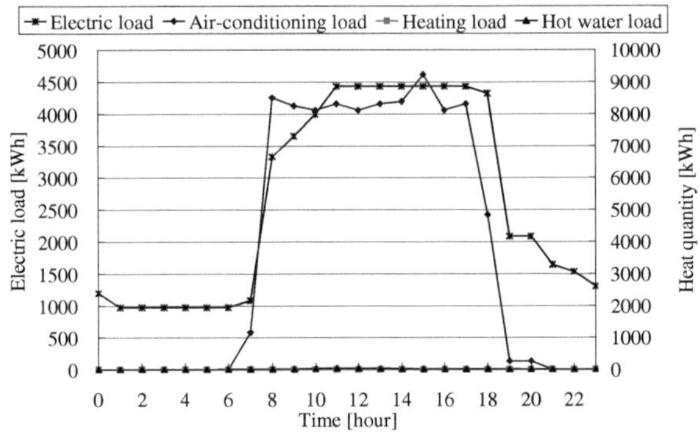

Fig. 4. Energy loads in summer season (August).

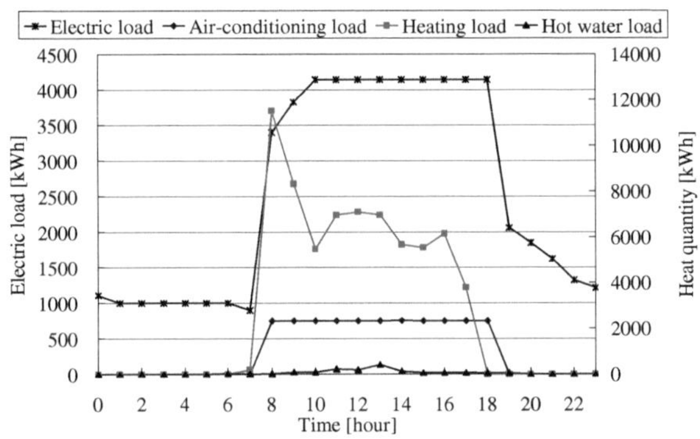

Fig. 5. Energy loads in winter season (February).

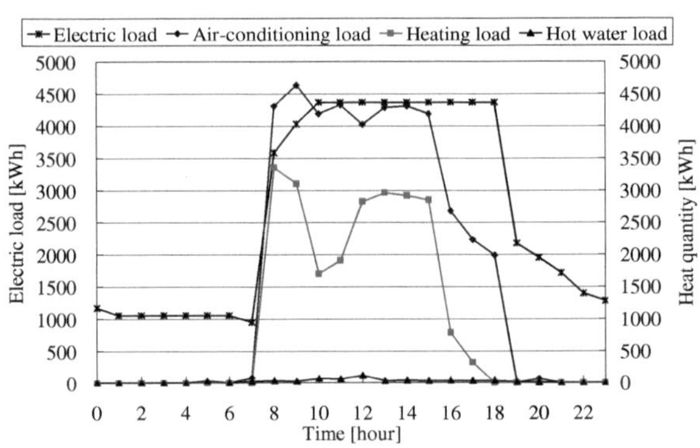

Fig. 6. Energy loads in intermediate season (November).

reduced compared with the conventional method. EPSO and APSO can generate better average results than the original PSO. Figure 7 shows examples of the detailed results by the three PSO based methods. Many heat energies are input to genelink, and heating and hot water loads are supplied by boilers using the original PSO method. On the contrary, many heat energies are input to heat exchangers for heating load and air-conditioning loads are supplied by fuel input in genelink by the evolutionary PSO methods.

5.2 An Automobile Company Factory

PSO based method has been actually installed in three factories of an automobile company in Japan [15]. Conventionally, the energy plants have been operated by

Table 1. Comparison of costs by the original PSO, EPSO, and APSO methods.

Method	Minimum	Average	Maximum
Conventional	100.0	-	-
Original PSO	98.68	98.80	98.86
Evolutionary PSO	97.96	97.97	98.00
Adaptive PSO	98.12	98.14	98.18

*) All of the value is the relative rate when the value of the conventional method is assumed to be 100.

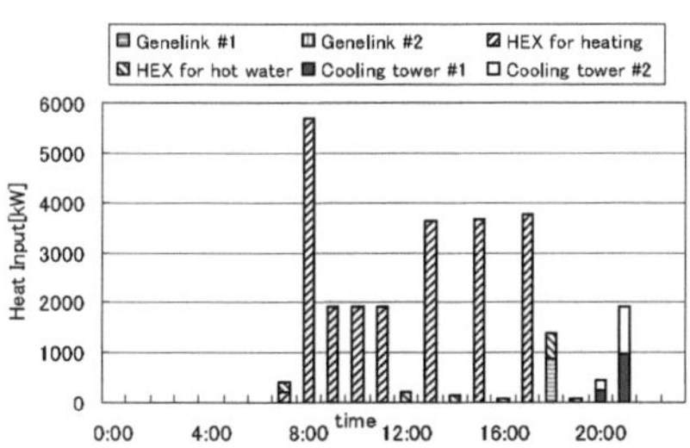

(a) The results of the EPSO method.

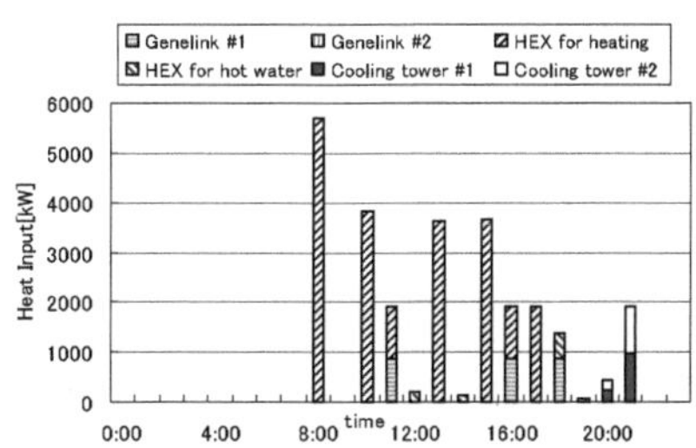

(b) The results of the APSO method.

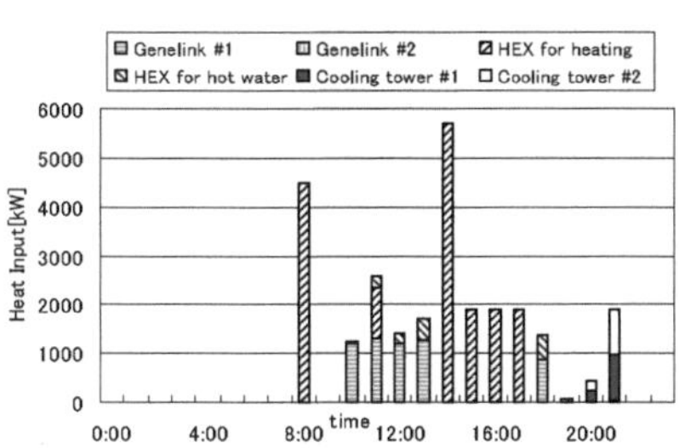

(c) The results of the original PSO.

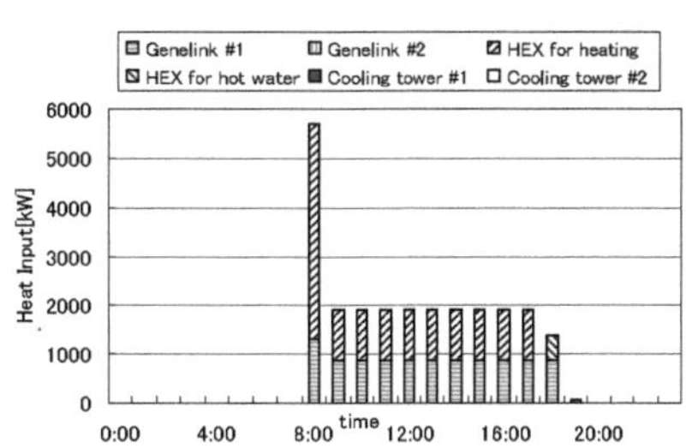

(d) The result of the conventional rule based method.

Fig. 7. Examples of the detailed planning results of heat energy distribution.

expertised operators using their knowledge considering weather forecast of temperature and humidity, heat storage volumes in heat storage tanks, and various operational conditions of factories (start and stop time of factory operation, overtime information, shift information, and rest information). The installed system forecasts various factory loads including electric, stream, and heat loads every 30 min. until 48 hours ahead. Weather forecasts are obtained every 3 hours from a weather forecasting company and they are utilized for load forecast. Load forecasting errors are within 3 % and high quality forecasting has been realized. Optimal planning is generated every 30 min. until 38 hours ahead and 10% energy reduction compared with operators' operation has been realized. Fig. 8 shows actual operation display hardcopies.

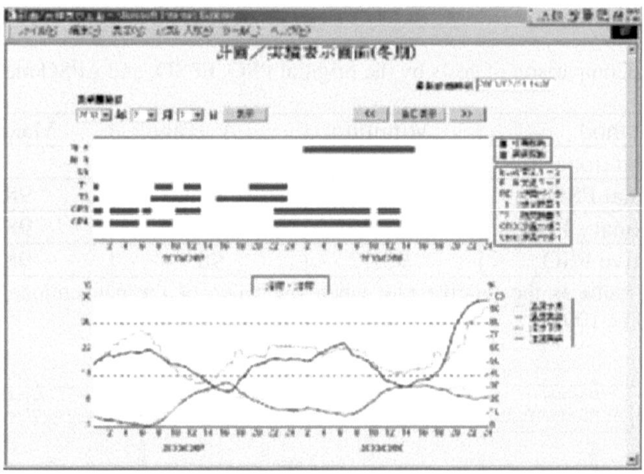

(a) Operational plans and results, and weather forecasts and results.

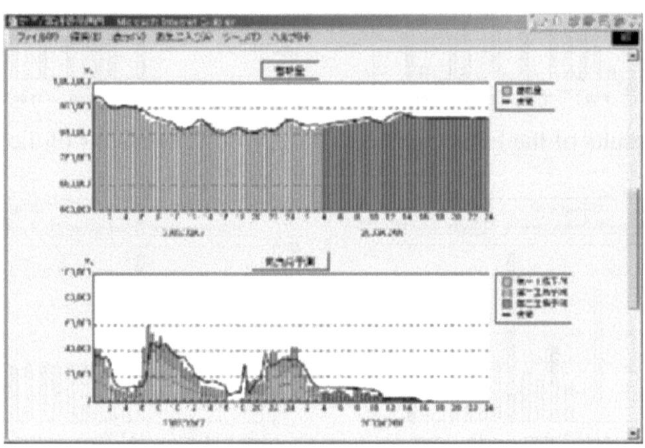

(b) Heat load forecasts and results

Fig. 8. Hard copies of actual operation display.

6 FeTOP- Energy Management System

An energy management system, called FeTOP, has been developed. It provides an optimal operational planning and control scheme of energy plants. It consists of three functions: an energy forecasting function, an optimal operational planning function of the facilities such as a cogeneration system, and a simulation function of the energy plant.

Figure 9 shows an example of practical system structure using FeTOP. The functions of FeTOP are installed in the PC servers. Data management and control signal output functions are installed in the Database & Control server. The load forecasting, the optimal planning, and the plant simulation functions are installed in the Planning server. The two servers and the process control system communicate through the local area network inside the factories and buildings. Since the forecasting results of weather conditions are necessary for the load forecasting functions, the weather forecast results are input to FeTOP from the weather information service provider. The planning results can be observed in the client PC installed in the energy management office by the operators. The operators can modify the next control signal if necessary.

FeTOP inputs measurement values of various sensors and consistency of the sensor information is important for realizing real optimal planning. The authors have developed the sensor diagnosis functions for FeTOP [17, 18]. The functions can find the sensors which should be repaired and modify the sensors' measurement values to the consistent values by the other sensors' measurement values. Using the functions, FeTOP can continue the optimal control even if some of the sensor measurement values are inconsistent.

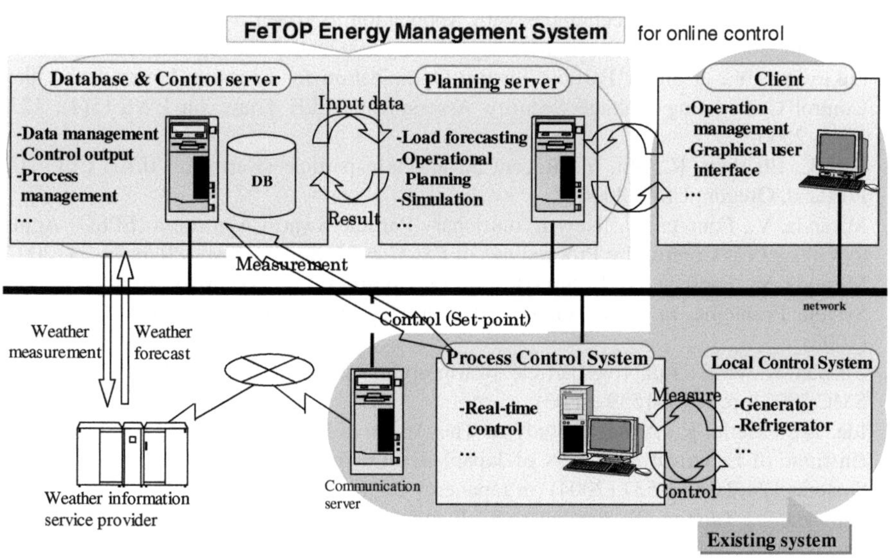

Fig. 9. An example of practical system structure.

7 Conclusions

This paper compares three PSO based methods for optimal operational planning of energy plants: the original PSO, the evolutionary PSO, and the adaptive PSO. The proposed methods are applied to operational planning of a typical energy plant and the simulation results indicate practical applicability of advanced particle swarm optimizations for the target problems. Following the comparison, an energy management system, called FeTOP, has been developed. FeTOP has been actually introduced and operated at three factories of one of the automobile companies in Japan and realized 10% energy reduction compared with operators' actual operation.

References

1. Ravn, H., et al.: Optimal scheduling of coproduction with a storage. Journal of engineering 22, 267–281 (1994)
2. Ito, K., Yokoyama, R., et al.: Optimal Operation of a Cogeneration Plant in Combination with Electric Heat Pumps. Transaction of the ASME 116, 56–64 (1994)
3. Yokoyama, R., Ito, K.: A Revised Decomposition Method for MILP Problems and Its Application to Operational Planning of Thermal Storage Systems. Journal of Energy Resources Technology 118, 277–284 (1996)
4. Kennedy, J., Eberhart, R.: Particle Swarm Optimization. In: Proceedings of IEEE International Conference on Neural Networks, Perth, Australia, vol. IV, pp. 1942–1948 (1995)
5. Kennedy, J., Eberhart, R.: Swarm Intelligence. Morgan Kaufmann Publishers, San Francisco (2001)
6. Fukuyama, Y.: Foundation of Particle Swarm Optimization. In: Tutorial text on Modern Heuristic Optimization Techniques with Application to Power Systems, IEEE Power Engineering Society Winter Power Meeting, January 2002, ch. 5 (2002)
7. Fukuyama, Y., et al.: A Particle Swarm Optimization for Reactive Power and Voltage Control Considering Voltage Security Assessment. IEEE Trans. on PWS 15(4), 1232–1239 (2000)
8. Hu, X., Eberhart, R., Shi, Y.: Recent advances in particle swarm. In: IEEE CEC 2004, Portland, Oregon, USA (2004)
9. Miranda, V., Fonseca, N.: New Evolutionary Particle Swarm Algorithm (EPSO) Applied to Voltage/Var Control. In: Proceedings of PSCC 2002, Sevilla, Spain, June 24-28 (2002)
10. Miranda, V., Fonseca, N.: EPSO -Best of Two World of Meat-heuristic Applied to Power System Problems. In: Proceedings of the 2002 Congress of Evolutionary Computation (2002)
11. Yasuda, K., et al.: Adaptive particle swarm optimization. In: Proc. of IEEE Int. Conf. on SMC 2003, pp. 1554–1559 (2003)
12. Ide, A., Yasuda, K.: A Basic Study of The Adaptive Particle Swarm Optimization. IEEJ (Institute of Electrical Engineers of Japan) Transactions on Electronics, Information and Systems 124(2), 550–557 (2004) (in Japanese)
13. Tsukada, T., Tamura, T., Kitagawa, S., Fukuyama, Y.: Optimal operational planning for cogeneration system using particle swarm optimization. In: Proc. of the IEEE SIS 2003, Indianapolis, Indiana, USA, pp. 138–143 (2003)
14. Kitagawa, S., et al.: FeTOP-Energy Management System: Optimal Operational Planning and Control of Energy Plants. In: Proc. of the IPEC-Niigata 2005, Niigata, Japan, pp. 1914–1920 (2005)

15. Fukuyama, Y., et al.: Optimal Operation of Energy Utility Equipment and its Application to a Practical System. Fuji Electric Journal 77(2) (2004) (in Japanese)
16. Fukuyama, Y., et al.: Intelligent Technology Application to Optimal Operation for Plant Utility Equipment. Journal of the Society of Plant Engineers Japan 14(3) (November 2002) (in Japanese)
17. Takayama, S., Fukuyama, Y., et al.: Sensor Diagnostic System for Plant Utility Equipment. In: Proc. of Annual Conference of IEE of Japan, No.4-180 (2004) (in Japanese)
18. Fukuyama, Y., Todaka, Y.: Energy Reduction of Utility Facilities Using Optimal Operation. Journal of Society of Instrument and Control Engineers, SICE (October 2006) (in Japanese)

10

Modelling Nanorobot Control Using Swarm Intelligence: A Pilot Study

Boonserm Kaewkamnerdpong[1] and Peter J. Bentley[2]

[1] Biological Engineering Programme,
King Mongkut's University of Technology Thonburi, Thailand
[2] Department of Computer Science,
University College London, UK

Abstract. Advances in the development of nanotechnology gradually bring the field into its next generation involving systems of nanosystems. These bring about opportunities for computer science researchers to contribute their work as guidelines for the realisation and development of nanorobot systems in the near future. It is anticipated that an early version of future nanorobots may potentially contain only essential characteristics and exhibit only simple behaviours. It is similar to social insects in nature; collaborative behaviour among such simple individual exhibits a remarkable degree of intelligence. Hence, swarm intelligence techniques inspired by social insects could potentially be applied for nanorobot control mechanism in self-assembly. This study models an early version of future nanorobots and a control mechanism using swarm intelligence, especially PPSO (the modification of PSO for physical applications), for self-assembly and self-repair to examine the minimal characteristics and functionality for future nanorobots.

Keywords: particle swarm optimisation, perceptive particle swarm optimisation, nanotechnology, nanorobotics, nanorobot, self-assembly.

1 Introduction

The United States' National Nanotechnology Initiative (NNI)[1], an organisation officially founded in 2001 to initiate the co-ordination among agencies of nanometre-scale science and technology, defines the term *nanotechnology* as

> Nanotechnology is the understanding and control of matter at dimensions of roughly 1 to 100 nanometres, where unique phenomena enable novel applications.

The field of nanotechnology emphasises in exploiting new phenomena and processes at nanometre scales and using atomic as well as molecular interactions to develop efficient manufacturing methods [1]. Nanotechnology has become

[1] See `http://www.nano.gov/html/facts/whatIsNano.html`

C.P. Lim et al. (Eds.): Innovations in Swarm Intelligence, SCI 248, pp. 175–214.
springerlink.com

public interest. In the last decade, investment in research has been dramatically increasing as the technology shows great promises to human society.

According to NNI, the development of nanotechnology has been divided into four generations [1]. The first generation, which ended in 2004, involved the development of *passive nanostructures* such as coatings, nanoparticles, nanostructured metals, polymers and ceramics. Nowadays, the products from nanotechnology are available at commercial level [2]. For example, nanoparticles are used to coat fibre, textiles, steel and so forth for stain repellent and corrosion protection; nanofilm or treated windshield repels rain and prevents sticking snow, ice, bugs and tar to increase driver vision. Coated nano-silver particles in wound dressing are used to improve antibacterial effectiveness [2]. In addition, nanoparticles are used widely in clear sunscreens and other deep-penetrating skincare products [3]. Furthermore, nanotubes are used to increase rigidity of a light-weight tennis racket [3]. The second generation involves the manufacture of *active nanostructures* including transistors, amplifiers, targeted drugs, actuators and adaptive structures should be achieved. Active nanostructures including a nano-motor and nano-valve are being developed [4,5].

Later, from the year 2010, nanotechnology should enter the third generation, *3-D nanosystems and systems of nanosystems*, for example: guided molecular assembling systems, 3D networking and new system architectures for nanosystems, robotics and supramolecular devices, and targeted cell therapy with nanodevices [1]. Finally, from the year 2020, the fourth generation of nanotechnology should be the generation of *heterogeneous molecular nanosystems*, which would focus on design and control of heterogeneous molecular systems as devices or components at atomic levels [1].

The development of nanotechnology involves interdisciplinary research. For many decades, it has been developed with cooperation from researchers in several fields of studies including physics, chemistry, biology, material science, engineering. Computer science has taken a role mostly in research tools, for example: a virtual-reality system coupled to scanning probe devices in nanomanipulator project, which allow us to investigate and manipulate the surface at atomic scales [6,7]. However, according to the NNI, the third and fourth generation of nanotechnology would rely heavily on research in computer science. Therefore, this brings about opportunities and challenges in computer science to design, program and control swarms of nanoscale robots in the near future [8,9].

With the inspiration of self-organised cooperation in social animals such as ants and termites, a machine learning algorithm using swarm intelligence has emerged in 1995 [10] and has been continuously developed since then. For nanotechnology agents, whose size is so small that we cannot see with our naked eyes and whose capacity and capability are limited, swarm intelligence is a possible solution [8]. With some modifications toward nanotechnology characteristics, these techniques can be applied to control a swarm of a trillion nanoassemblers or nanorobots (once realised).

This chapter introduces the pilot study of using swarm intelligence techniques to control a swarm of modelled nanorobots for self-assembly and self-repair tasks according to the concept of nanotechnology. We discuss the opportunities to get involved in nanotechnology, especially nanorobotics in section 2. The nanorobot swarm system—or NARSS—built to model a swarm system of early version of future nanorobots and its swarm-based control is described in section 3. Section 4 investigates the performance of NARSS in single-layered surface coatings.

2 Getting Involved in the Third Generation of Nanotechnology

The idea of nanotechnology has been introduced and developed for many decades. It was the visionary talk entitled "There's Plenty of Room at the Bottom" [11] given by Richard Feynman in 1959 that inspired many researchers in physics, material science, chemistry, biology and engineering to become nanotechnologists. Feynman suggested a new method of production by rearranging the atoms with precision, using several tiny agents working in parallel. This is later referred as *bottom-up technology*, an alternative style of technology to *bulk* or *top-down technology* [12] that we have been using to construct a refined product from bulk materials since our ancestors.

Later in 1986, Drexler described in his book "Engines of Creation" [12] in more details of how nanotechnology would advance in the future. Drexler described possible benefits that can arise from nanotechnology once realised, for example healing, thinking machines, self-repair machines and space exploration. Drexler predicted that we would build mechanical components in molecular size and assemble those components into nanomachines. He expressed that these early version of nanomachines could be living ones, which could build a better, more complex version of themselves. Like enzymes, these molecular machines would have the ability to bond and detach molecules. Like ribosomes, the natural assemblers in cells, nanomachines would have programmability. Drexler referred these nanomachines that can be programmed to join or split molecules as the second-generation nanomachines or *assemblers*. While assemblers would allow engineers to synthesise objects, another type of nanomachines, *disassemblers*, that are able to break bonds from layer to layer, would help scientists to study and analyse any unknown objects. Moreover, disassemblers could break molecules for assemblers to use as raw materials. Since a great number of assemblers would be required to build a product, the idea of replicating assemblers that copy themselves like viruses to build other products arose. Ultimately, assemblers would be able to build anything from common materials using clean and economical manufacturing process. Drexler seemed convinced that the emergence of such nanomachines would be virtually inevitable. In 1992 [13], he conducted a detailed technical analysis on molecular manufacturing that such molecular manufacturing systems are feasible.

2.1 Nanorobots

In the current development of nanotechnology toward the realisation of nanoassemblers, referred as nanorobots or nanobots, there are two approaches: mechanical and biological approach. In mechanical approach, the vision of early nanorobots is a nanometre-scale device comprising of nanomechanical parts—i.e. motors, actuators and sensors—with robotic arms to manipulate atoms in place and nanocomputers to control the behaviour [12]. Drexler conducted an analysis on the feasibility of molecular manufacturing including nanoscale components and their dynamics in [13]. Following the mechanical approach, Kelly and his group developed a single-molecule *nanocar* consisting of a chassis, axles and four buckyball wheels [14]. The nanocar is measured only 3 to 4 nanometres across. It functions similarly to conventional vehicles by rolling its wheels perpendicular to its axles to move on a surface, but nanocar manipulation requires external stimuli. The nanocar's buckyball wheels roll on the heated gold surface between 170 and 225 degree Celsius; at such high temperature, the adhesion force between the fullerene wheels and the gold surface is reduced. The movement of the nanocar can be directed by electrostatics from the STM tip. The group aims to subsequently develop nanoscale transporters that could convey atoms and molecules in non-living fabrication environments similar to hemoglobin conveying oxygen to living cells[2]. Even though the nanocar can move (with the help from users), it has yet no ability to manipulate atoms or molecules; hence, the path to the generation of nanorobots is still distant.

The study of similarly-sized biological machines—organic cells—suggests that there may be more effective alternatives to mechanical ones. In 2004, Liao and Seeman [15] were able to develop a DNA nanomechanical device that mimics the ribosome transitional capabilities to enable positional synthesis DNA molecules into a specific DNA sequence. This device is potentially used in information encryption and a variable-input device for DNA computers [15]. Meanwhile, Sherman and Seeman [16] demonstrate molecular walking biped DNA robot moving forward and backward along a defined track. This DNA-based nanodevice system consists of two components: a triple-crossover molecule footpath and a designed DNA-strand biped. Synthesised fuel set and unset strands are introduced into the solution. When the biped stands on the footpath, a set strand hybridises each of the single-strand biped. For the biped to take a step, the system requires separate annealing for the biped and the footpath. An unset strand binds and removes one of the set strand attached to the biped; hence, the foot is released from the foothold of the path. The free foot can be set on another foothold. Then, the same process applies for another foot and so on. The biped could be used to transport loads or to represent the state of a DNA-computational machine when operating a 2D footpath. In the future, the structural properties in nucleic acids—i.e. DNA— could be used for other purposes including nanorobots [17]. Scientists are investigating nucleic-acid nanotechnology for self-assembly and other applications[3].

2.2 System of Nanorobots and Nanorobot Control

Based on either mechanical or biological approach, the ideal version of nanorobots to achieve massively parallel self-assembly and other tasks has not yet been

[2] See `http://www.rice.edu/media/nanocar.html`

[3] See `http://www.spacedaily.com/news/spacemedicine-05f.html`

realised. Nevertheless, Cavalcanti and Freitas [18,19] developed a real-time simulation of mobile robots at nanoscales to perform biomolecular assembly manipulation for nanomedicine, which will potentially be the first application for nanorobots. The nanorobots assemble molecules and deliver assembled biomolecular substances to a predefined set of organ inlets to improve the nutritional state [19]. The nanorobots are designed to operate in aqueous environment with six degree-of-freedom movement [19]. Each nanorobot has

1. molecular sorting rotors and a telescoping-manipulator robotic arm for manipulation of molecules,
2. three fins and bi-directional propellers for navigation,
3. a receiver for macrotransponder navigation system,
4. sonar sensors for collision and obstacle identification,
5. acoustic communication sensors for explicit communication with others, and
6. diamondoid exterior that is biocompatible to human body.

Using virtual reality, the simulation simulates nanorobot locomotion and kinetics, which is based on the concept of underwater robotics, for rigid body motion in hydrodynamics with low Reynolds number [19]. The positions and trajectories of molecules are randomly generated. Cavalcanti and Freitas use a macrotransponder navigation system for positioning nanorobots to the target organ inlets with accuracy [19]. An external signal generator, which is located on the skin near the target area, is required. The receiver must be mounted on each nanorobot. The type of such signal generators and receivers is not specified. Nanorobots move toward the target area and avoid obstacles detected by their local perception with sonar sensors. When a nanorobot finds nutrition molecules, it manipulates the molecules with its robotic arm and delivers to organ inlets. When the delivery is completed, a nanorobot communicates a confirmation signal to its partner. Unnecessary explicit communication among nanorobots is limited in order to minimise the energy consumption.

The future nanorobots must tackle the problem of massively parallel automation control. Cavalcanti and Freitas applied a genetic algorithm for decision control. Like social insects, nanorobots act according to a set of interaction rules in conjunction with the evolutionary decision resulting from processing the perceived signal [18]. The nanorobot decision is represented by a chromosome-like message describing how, when and to which organ inlets it moves. One strategy is that nanorobots compete against each other to improve organism health and, meanwhile, prevent the repeat of nutrition delivery to the same organism [20]. Alternatively, nanorobots can work in pairs and take turns to feed a specified number of organ inlets to avoid overdose [19]. For motion control, a stochastic feedforward artificial neural network is used for finding the optimal nanorobot path and dynamic obstacle avoidance as it requires low computational effort [18]. If a nanorobot is ready to deliver to the organ inlets corresponding to the evolutionary decision, it follows the delivery route to visit specified organ inlets. Otherwise, it follows the verification route to visit and check the nutrition levels in other organ inlets. In simulation, such nanorobot system accomplished its task in a dynamic environment [19].

3 Nanorobot Swarm System

As nanorobots have not yet been developed, studies on nanorobot modelling and simulation are important to the advancement of nanotechnology. One reason is that the study demonstrates the possibility of nanorobot systems with specified characteristics to accomplish its tasks; if the systems succeed, those nanorobot characteristics can serve as guidelines to develop future nanorobots. Like in the field of robotics, the better the simulation can simulate the real situations at nanoscales, the more accurate the result of the designed nanorobot systems will be. Since not all molecular phenomena at nanoscales have been identified, the simulation must be continually modified according to latest molecular studies.

Cavalcanti and Freitas' version of nanorobots is based on mechanical approach. These nanorobots require sophisticated molecular manipulation tools, sensors and signal generators for explicit communication. It can be considered as an advanced version of nanorobots. For nanoscale robots, it is possible that each of which comprises of limited capabilities due to small size. The early-stage nanorobots are probably composed of only essential robotic attributes. In 1996, Holland and Melhuish [21] investigated the abilities of single and multiple agents on a task with programmable agents with limited sensing, mobility and computation (similar circumstances as future nanorobots). The task to be solved by the agents in their studies was to learn to move toward a light source by using simple rule-based algorithms. In the case of single agents, the result was efficient, but performance degraded as the amount of noise increased. In the case of multiple agents, the best result was from the algorithm that formed collective behaviours akin to genuine social insects. This investigation showed that emergent collective intelligence from social interactions among agents modelled on social insects could cope with the limited capabilities that would be inevitable in future nanoscale robots. Such intelligence in social insects, well-known as *swarm intelligence*, will be critical for the future of bottom-up nanotechnology.

Inspired by collective intelligence in social insects each of which exhibits simple behaviours, swarm intelligence techniques are developed and applied for solving optimisation problems [22,23,24]. Unlike typical optimisation problems, the applications for which the self-assembly and self-repair swarm system is designed are not one optimal points but a series of optimal points. The swarm system must be designed so that individuals in the swarm cooperate and together arrange themselves into structures. In addition, they are capable of repairing the structure when damaged as long as they are in the operating environment. Such a system is essential for future nanotechnology to improve our ways of life. Though the fundamental sciences for developing advanced machines in nanoscales are not available at the present time, this study aims to demonstrate that a swarm system of nanorobots with some essential characteristics can be functional in nanotechnology from a computer science point of view.

3.1 Self-assembly and Self-repair

Whitesides and Grzybowski [25] gave a definition of self-assembly as

> Self-assembly is the autonomous organization of components into patterns or structures without human intervention.

Self-assembling processes are commonly found in nature; for example, after a small nucleus of crystal is formed, crystal grows by the addition of atoms to the face with greatest surface tension and forms into shapes such as cube and octahedra [26]. Self-assembly can be classified into two types: static and dynamic self-assembly [25]. The systems with static self-assembling process do not dissipate energy, whereas the dynamic self-assembly systems dissipate energy during the formation of structures. Most research in molecular self-assembly involves the static one [25]. Dynamic self-assembly is complex and often found in biological systems. Nature has shown this alternative assembly method to directed–assembly techniques which require careful fabrication with precision control that we have been using in production. The revolutionary self-assembly, however, requires more restrictive design than directed methods [27]. Goodsell described some fundamental points in [27] as follows:

1. As systems of identical modules carrying limited information can create large assemblies like bricks constructed into a wall, modularity can bring several advantages to self-assembly.
2. Symmetrical modules can be used to control the size of the assemblies, whereas quasi-symmetrical modules[4] are used for larger assemblies that are too large to build with symmetrical modules.
3. Modules must interact uniquely with other modules in the same assembly and interact differently with modules from other assemblies.
4. To form an assembly, modules must assemble in specific orientations.
5. In crowded environments, the distance among modules decreases and the association of molecules into larger assemblies increases.
6. No information to specifically guide the assembly is required.

Self-assembly is similar to self-organisation in the aspect that they both produce patterns. Their processes are, however, different. The crucial distinction between self-assembly and self-organisation is the initial differences and/or relationships among modules or components [28]; the components in self-assembly encode information on characteristics which determine the interactions among components [25], while in self-organisation such encoding is not required [28]. In other words, self-organisation has non-specific surfaces of interaction and allows more interactions among neighbouring modules [27]. Thus, self-organisation is a mechanism for creating flexible and self-repairing structures [27].

Self-assembly through a self-organisation mechanism is effectively simpler, more robust, less prone to failure and more capable to self-repair [28]. Sendova-Franks and Franks gave an example of sorting the series of integers from 1 to

[4] Quasi-symmetrical modules are built by placing multiple identical molecules at each symmetrical position [27].

100 in [28]. In self-assembly with explicit coding, the module carrying number 50 could be instructed to position on the right of module 49 or on the left of module 51. In self-assembly through self-organisation, each module might be regulated by a simple rule that the module with a larger number must be on the right of the module with a smaller number. In the case of self-assembly with explicit coding, the global structure will not emerge if a single module carrying an intermediate number is missing [28]. On the contrary, missing number is not a problem for self-assembly through self-organisation [28].

An example of self-assembly through self-organisation in biological systems is lipids self-assembling into membranes. Lipids are small molecules composed of soluble chemical substances in water and connected to insoluble hydrocarbon tails. They have been used by living cells for infrastructure [27]. Lipids connect to each other so that the hydrophobic segments are shielded from water when they are in water with higher concentration than their critical concentration. Cylindrical-shaped lipid molecules self-assemble into bilayer membranes to resist molecules and ions from passing through. This is driven by self-organisation with the need to shelter the hydrocarbon tails from water [27].

In [12], Drexler suggested a system comprising of nanoassemblers mechanically arranging atoms or molecules together with manipulator arms operating in parallel. Nevertheless, with the current technology in nanomanipulators this process will be time-consuming for mass production. A more effective way is perhaps self-assembly through self-organisation, which has shown astonishing capabilities in biological systems.

3.2 Characteristics of Nanorobots

In this study, self-assembly through self-organisation is adopted for NARSS. The early-stage nanorobots will probably have limited capabilities and exhibit only simple actions as insects do. Instead of playing the role of assemblers that manipulate molecules into structures as in Drexler's early version of nanoassemblers [12], nanorobots in this system may not have robotic arms to manipulate and assemble molecules into a fine product but rather act as assembly modules and become incorporated in final structures. In order to fulfil self-assembly process, each nanorobot must be able to

1. move around the environment,
2. interact with other nanorobots,
3. carry defined characteristics for the assembly task, and
4. connect to another nanorobot at a specific surface.

Similar to macroscale robots, nanorobots will require features including actuator, signal generator and sensor, programmability, and connection for self-assembly. The following sections describe these features, examples in biological systems and their potential from the literature. Note that energy is a fundamental to nanorobots and is not explored in this study.

Actuator: As both self-assembly and self-organisation are motivated by multiple interactions among entities, the change of nanorobot positions will encourage interactions with other nanorobots. A dynamic environment may cause the position of nanorobots continually change even though the nanorobots do not move; however, with increased nanorobot movement the chance of engaging in interactions with other individuals can be increased.

Many bacteria, such as *Escherchia coli* or *E. coli*, use multiple flagella. Each of which has a small rotary motor to induce their locomotion [29]. When the motors all turn counter-clockwise, the cell moves forward; this movement is known as *bundling*. When the motors turn clockwise, the movement of the cell changes in random direction; this movement is called *tumbling*. Bundling and tumbling allows the cell to move toward the direction that yields more favourable conditions for survival [29].

Nanoscale rotary motors have already existed in nature; ATP synthase is an enzyme that synthesises adenosine triphosphate (ATP) which is the universal energy carrier in all living organisms[5]. ATP synthase is a combination of two motors known as F_0 and F_1. The F_0 motor is situated in the membrane, whereas the F_1 is above the membrane. The F_0 motor is driven by the flow of protons or ions through the motor [27]. The rotor portion of the F_0 motor has an eccentric axle that is connected to the F_1 motor [27]. The F_1 motor is driven by the cleavage of ATP in the motor [27]. To generate ATP, the F_0 motor rotates and forces the F_1 rotation to increase the affinity for combining adenosine diphosphate (ADP) with phosphate to form ATP. Then, with further rotation ATP is released into the organism. The ATP synthase enzyme has been applied for rotary biomolecular motor-powered nanodevices; Soong *et al.* [30] attached ATP synthase to nanopropellers (150 nm in diameter and 750 to 1400 nm in length). The biomolecular motor can drive nanopropellers with mean rotation velocity of 8.0 ± 0.4 and 1.1 ± 0.1 rps for nanopropeller length of 750 and 1400 nm respectively [30].

Signal Generator and Sensor: In the early version of future nanorobots, interactions in the form of direct communication may not be possible; each nanorobots may perceive other nanorobots within their proximity and act according to interaction rules as programmed. If this is the case, each nanorobots must be able to sense the presence of one another from either the properties of nanorobots or signals generated by nanorobots to establish an interaction between two nanorobots.

An example of sensing in nature is bacteria seeking out food by sensing nutrient levels. They sense the environment at different times which are then compared to determine whether the nutrient levels are decreasing or increasing. If the nutrition is increasing, the cell bundles in that direction; if otherwise, the cell tumbles by reversing flagellar motors to a different direction. This method leads to a random motion that allows bacteria to survive in their environment [27].

[5] See `http://nobelprize.org/nobel_prizes/chemistry/laureates/1997/press.html` for more information.

Nevertheless, bacteria only have the information of the nutrient concentration at their current positions in comparison to the previous positions; there is no mechanism to determine the best direction to travel according to their experience. Nevertheless, bacteria can communicate to one another through the mechanism called *quorum sensing*. Bacteria monitor their population density by the production of a low-molecular-mass signal molecule, called *autoinducer* or *quormon* [31]. For example, N-acyl-homoserine lactone (AHL) signal is used in cell-cell communication in gram-negative bacteria, such as *Vibrio fischeri*[6] [31]. Two components of V. fischeri involves in the AHL signalling system includes *luxI* synthase gene and *luxR* code. The AHL signal molecule is produced by *luxI* and diffuses across cell membrane to the environment. The greater the population density, the higher the AHL concentration. When the concentration reaches a threshold, the AHL signal molecule from the environment diffusing into the cell binds to the *luxR* transcriptional activator. The resulting molecule binds to the bioluminescence operon *luxICDABEG* and, bioluminescence is induced [31].

In nature, bioluminescence is used for attraction purposes by several animals; for example, deep sea fish such as the anglerfish lure their prey with luminous rods [32]. The North American fireflies, *Photinus pyralis*, flash light to communicate and attract mates [33]; the female fireflies wait for the males to flash and respond with a different frequency flash when a flash from the males is perceived [33]. The same mechanism can be adopt for communication among future nanorobots. Bioluminescence is a result of a chemical reaction called *chemiluminescence* that converts chemical energy to radiant energy [33]. To resemble bioluminescence communication, quantum dots can be applied. Quantum dots have a great potential as biological markers [34,35,36]. Semiconductor quantum dots have the ability to emit fluorescent in a range of colours. By varying the size of such spherical particles (typically from 1 to 12 nm [34]), the frequency of emitting light—hence, colour—is altered. These nanocrystals can be attached to biomolecules to replace traditional fluorescent probes in order to visualise cells [37] and could be used to study the pathogenesis of vascular diseases [35].

Moreover, quantum dots also have a potential in the use as optical sensors [34]. The studies revealed that chemical or physical interactions with its surface state can affect the sensitivity in the luminescence of quantum dots [34]; it is expected that the interactions between the analyte and the surface of quantum dots would change the surface charges and, hence, photoluminescence emission [34]. Optical sensors based on the interaction on the quantum dots' surface have been exploited [34]. Such method could be applied for sensing units in future nanorobots.

Programmability: To self-assemble into a desired structure, each nanorobot must carry defined characteristics specified for the assembly task. In self-assembly, the interaction among components will be determined by these characteristics. On the other hand, less specific characteristics of nanorobots can be used in self-assembly through self-organisation; however, each nanorobot must contain

[6] V. fischeri are marine bacteria that become bioluminescent when the population density is high.

interaction rules to regulate the interaction with another nanorobot to flexibly create desired structures. In the early stage of nanorobotics, nanorobots are simple and may not require to modify the program during their operation; hence, each nanorobot may be programmed at construction.

In biological systems, cells contain molecular tools for programmability; the existing nanoassemblers, *ribosomes*, build proteins according to a set of instructions stored in ribonucleic acid (RNA) [27]. The natural blueprint for building proteins is encoded in DNA strands. The information in DNA is transcribed into RNA. Then, ribosomes translate the code in RNA and connect proper amino acids into a protein chain [27]. DNA can be considered as archive of genetic information. Information is copied onto RNA which is used as templates for constructing proteins in ribosomes; this is similar to paper tapes or punch cards that had been used to store information for processing in the early version of computers.

Apart from its programmability through RNA in biology, DNA has several advantages for nanotechnology. DNA molecules have been constructed into DNA tiles with overhanging sticky ends and used as building blocks to create patterns for RAM units with demultiplexed addressing [38]. Such self-assembly by DNA tiles can be used in other nanomechanical devices [39]. Similar method using designed DNA sequence with sticky ends has been applied for a molecular DNA computer [40]. This biomolecular nanocomputer can function as disease indicators [40]. In 2004, it has been demonstrated on drug administration for the prostate cancer [41]. Thus, DNA could enable programmability in nanorobots.

Connection: In order to self-assemble into a structure, each nanorobot must be able to form a connection with one another. In biological systems, atoms can connect to one another with several types of interactions: for example, covalent bonds, hydrogen bonds, dispersion-repulsion forces and electrostatic forces. *Covalent bonds* are the strongest interactions within biological molecules [27]. These bonds are stable and highly directional. Atoms connecting through covalent bonds align in rigid, defined geometry. To create or break such bonds, a significant amount of energy is required [27]. Another kind of atomic bonds is *hydrogen bonds* formed between a hydrogen atom and another oxygen, nitrogen, or sulfur atom. Compared with covalent bonds, hydrogen bonds are weaker and less directional but still stable in biological contexts as they are stronger than typical thermal energy [27]. Because oxygen and nitrogen atoms are abundant in biological molecules, hydrogen bonds are widely used within bionanomachines [27].

Non-bonded forces dominate the interactions within molecules and between molecules. Two opposing forces, *dispersion-repulsion forces*, defines the space between two atoms. Dispersion forces draw neighbouring atoms together with a weakly favourable interaction, whereas repulsion forces keep atoms from interpenetrating each other. These forces are required in bionanomachinery at molecular level for the stability of bionanomachines [27]. Another type of non-bonded interaction is *electrostatic forces*. These forces are long-range, non-directional interactions between atoms with electronic charges. The opposite charges on atoms attract to one another, whereas the identical charges repel one another. Apart from these interactions, other properties such as hydrophobic effect can dominate the properties and interactions of bionanomachines [27]; the self-assembly of cylindrical-shaped lipid molecules into bilayer membranes is an example.

An example of connections in current nanotehcnology development is DNA sticky ends. These sticky ends can be created by restriction endonucleases, which is an enzyme that cuts double-stranded DNA. DNA sticky ends can be linked together by DNA ligase when both ends are complementary. Nadrian Seeman has pioneered the construction of multi-dimension structures using DNA with stricky ends [27]. Double-stranded DNA with sticky ends have been used to construct closed polyhedra, such as cubes and truncated octahedra, as well as DNA tiles with the ultimate aim to develop biochip computers and nanorobotics[7].

3.3 Nanorobot Control

According to the potential characteristics of the early stage of future nanorobots in the previous section, each nanorobot in NARSS is designed to hold the following features:

1. *actuator* :- each nanorobot has actuator units that allow it to move in a direction with controlled velocity. More than one motor may be allocated around the nanorobot to expand the range of different directions and velocities.
2. *signal generator and sensor* :- each nanorobot interacts with one another through signalling and sensing units. Each nanorobot is designed with the ability to generate signal which can be chemical substance, magnetic force, luminescent substance or other suitable means. The signal presents a strong intensity near the nanorobot. As further away from the nanorobot, the signal is gradually reduced. Each nanorobot has sensing units that complement the signalling units to perceive signals from other individuals and the environment. The communication channel that allows two nanorobots exchange information is not included.
3. *programmability* :- each nanorobot is required to be programmed with a set of interaction rules comprising with simple mathematical calculations to govern the movement according to nanorobot control mechanism. Some factors in the conditions for these rules can be adjusted for effective control. A data storage that can hold values of factors for a period of time is required. Such storage may be in biomolecular version of electronic flip flops, i.e., genetic flip flop in E. coli [42] and biological integrated circuit [43].
4. *connection* :- each nanorobot is designed to allow a definite number of possible connections to other nanorobots. Six connections aligning in a hexahedron—six-faced polyhedron—with four degree sequences equally in each face are used and illustrated in Fig. 1.

Two main types of swarm intelligence techniques include *ant colony optimisation* (ACO) and *particle swarm optimisation* (PSO). Even though ACO is powerful and can be applied to both discrete and continuous optimisation problems, they have a limitation to employ in physical applications. As pheromone is

[7] See more information at `http://seemanlab4.chem.nyu.edu/`

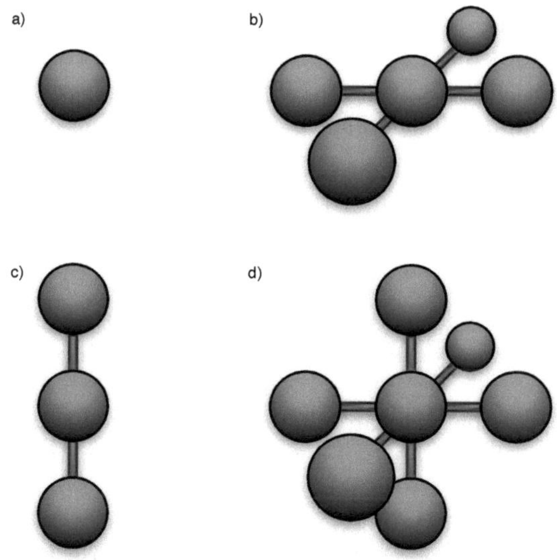

Fig. 1. The connection between nanorobots: a) a single nanorobot, b) nanorobots with horizontal connections, c) nanorobots with vertical connections and d) nanorobots with both horizontal and vertical connections

the essence of ACO meta-heuristic, there must be pheromone (or other similar-purpose substances) and the environment to which pheromone is deposited. Payton *et al.* [44] simulate virtual pheromone using infrared transceiver mounted on physical robots called *pheromone robots* or *pherobots*. Instead of laying chemical substance, a pherobot transmits message of virtual pheromone to neighbours which determine pheromone intensity from the signal and the estimated distance from the sender. Then, pherobots relay the received message to their neighbours. The message is relayed for a number of times specified by the sender. This method can crudely simulate pheromone detection and its gradient: as ants lay pheromone on their path, only ants in the proximity of the path can detect pheromone trail and be influenced. Nevertheless, such virtual pheromone requires message transmission which may not be plausible in early stage of nanorobots.

Using similar representation to physical agents, PSO seems plausible to apply in physical applications including nanorobot coordination control. The inexpensive requirement in memory and computation suits well with nano-sized autonomous agents whose capabilities may be limited by their size. Nevertheless, the conventional PSO algorithm requires complex, direct communication among particles in the neighbourhood which might not be possible by real nanorobots. To apply PSO for nanorobot control, a modification of the algorithm is required. In [45,46], the perceptive particle swarm optimisation (PPSO) algorithm, which is a modification of the conventional PSO algorithm, is proposed to mimic behaviours of social animals more closely through both social interaction and environmental interaction for physical applications. In the experiments in function optimisation problems, PPSO performed well when compared with the conventional PSO algorithm despite its restrictions on communication [46]. PPSO can potentially be applied for nanorobot control in self-assembly and self-repair.

PPSO as Nanorobot Control: PPSO is relatively similar to the conventional PSO algorithm except for two parts:

1) To substitute for the lack of direct communication among particles in the swarm, PPSO is assumed that each particle has sensory units to sense the presence of its neighbours and the search space, which is the environment of the swarm. The sensory units can be a CCD camera, infrared sensors, chemical-detected sensors, or other sensors that are appropriate for the applications. Each individual has a finite range of perception according to the operating range of sensors. Instead of directly exchanging information among particles in their neighbourhoods, each particle perceives the positions of other individuals and the search space within its perception range as social insects observe other individuals and the world through senses. PPSO particles are attracted to the better positions in the search space as well as to the neighbours they perceive.

2) For n-dimensional optimisation problems, PPSO operates in $(n + 1)$-dimensional search space instead of operating in n-dimensional search space as in the convention PSO algorithm. The added dimension represents the underlying performance of particles at their positions in n-dimensional space. Note that the exact performance at a specific position in the space is unknown to the particles in PPSO. Adding the additional dimension and the ability to observe the search space allows particles to perceive their approximate performance. Consider an n-dimensional function optimisation problem. In the conventional PSO algorithm, particles fly around the n-dimensional space and search for the position giving the greatest performance measured by using the function to optimise. On the other hand, PPSO particles fly around $(n+1)$-dimensional space to observe the space and find the optima of the landscape; an n-dimensional objective function for the function optimisation problem is treated as an $(n+1)$-dimensional physical landscape (the extra dimension being fitness), which particles are constrained to fly above and never below. Because PPSO particles can fly over a physical fitness landscape with discontinuities and noise, they can observe landscape's peaks as well as other particles positions and endeavour from afar to find a good solution. Fig. 2 illustrates the difference between particles in the conventional PSO and PPSO algorithms in function optimisation problems.

The perception of each particle in PPSO simulates that of an physical agent in physical applications where communication and information sharing within the swarm may be not available. Compared with the conventional PSO algorithm, the simulated version of interactions among individuals through perception for physical applications in PPSO leads to the degradation in the quality of social interaction as there is no information sharing among particles; hence, each particle must rely on its perception of the environment and influence from its neighbour randomly chosen from its neighbourhood. Based on four fundamental

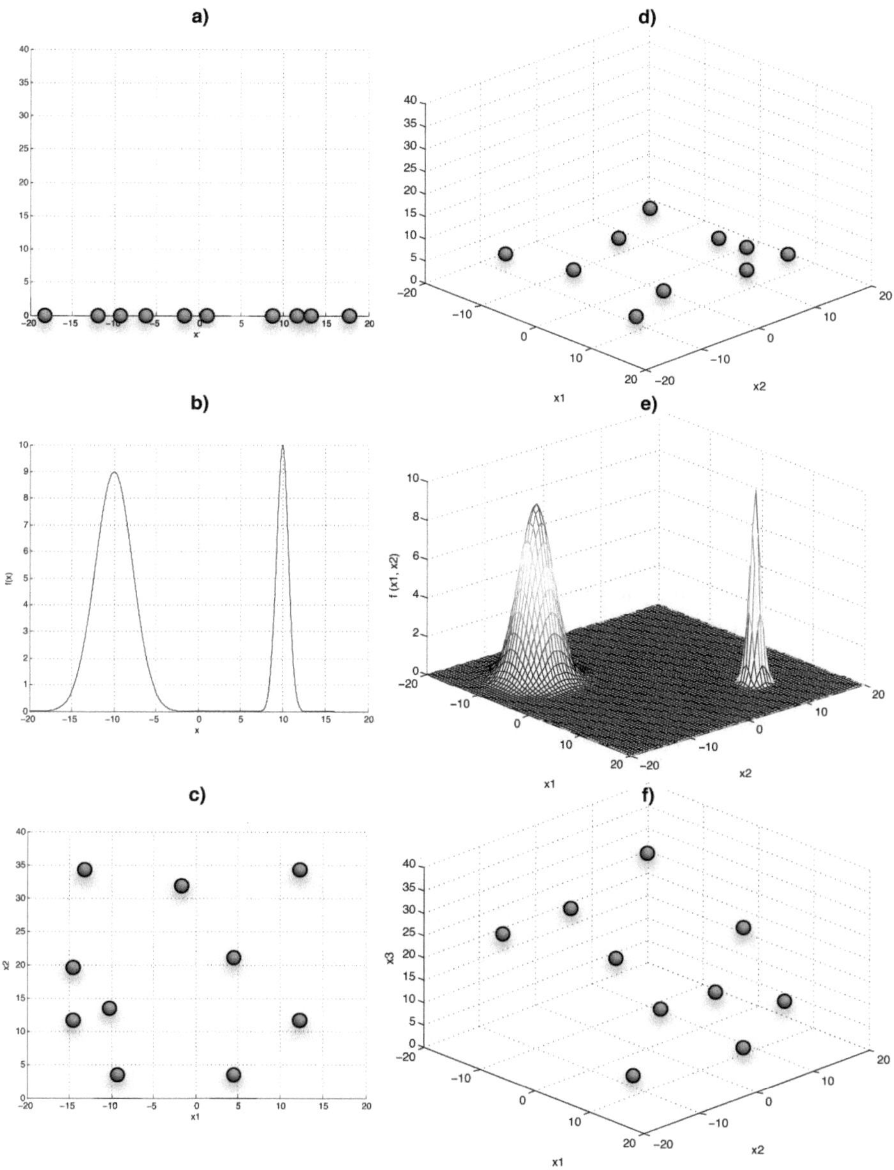

Fig. 2. The comparison between the conventional PSO algorithm and the PPSO algorithm in function optimisation problems: a) PSO particles represented by green dots operate in a one-dimensional problem where its objective function is shown in b); c) PPSO is employed in the same problem which is treated as a landscape optimisation problem; d) PSO particles operate in a two-dimensional problem where its objective function is shown in e); f) PPSO is employed in the same problem.

components of self-organisation (including positive feedback, negative feedback, amplification, and multiple interaction) in [47], the form of feedback in this case becomes obscure. In social insects such as ants, feedback is given in the form of pheromone trail leading ants to the food source. In the conventional PSO algorithm, particles determine the feedback from their positions and share the most positive feedback with their neighbours. Without the ability to share information, particles in PPSO cannot explicitly offer their findings to other individuals; however, particles can still influence their neighbours which implicitly give an indirect form of feedback. The randomness in the selection of influencing neighbours promotes the exploration for the better solutions in the search space. Experimental results in [46] has shown that particles with the restrictions on communication capabilities can collectively behave as a swarm and find good solutions for experimented function optimisation problems.

For a swarm of m nanorobots, let $\mathbf{x}_i(t)$ denotes the position of nanorobot i in the search space at time step t, where $\mathbf{x}_i(t) \in \Re^{n+1}$. The initial positions of nanorobots are distributed over the search space; the search space, however, includes an additional dimension for fitness landscape with values in the range from zero to $x_{max,n+1}$. The velocity with which nanorobot i travels in the space at time step t is denoted as $\mathbf{v}_i(t)$, where $\mathbf{v}_i(t) \in \Re^{n+1}$. For each dimension, the initial velocity of each nanorobot is randomly generated between $-V_{max}$ and V_{max}. The magnitude of nanorobot velocity should not exceed V_{max}. Thus, the initial velocity is regulated by the rule:

$$\texttt{if } \left|\mathbf{v}_i(0)\right| > V_{max} \texttt{ then } \mathbf{v}_i(0) = V_{max}\left(\frac{\mathbf{v}_i(0)}{\left|\mathbf{v}_i(0)\right|}\right). \tag{1}$$

At initial positions, the local axes of nanorobots are aligned with the world axes. Nanorobots maintain their main axes (for example $+x$ axis) in the directions of the velocities they travel without tilting. To determine the orientation of local axes with respect to the world axes in three-dimension space, nanorobots calculate the Euler angles of their velocity vectors, $\mathbf{v}(t)$; as nanorobots do not tilt, only the right-handed rotation angles about y- and z-axis (referred as pitch and yaw angles respectively [48]) are determined as shown in Fig. 3.

For simplicity and reduction of computation time in simulation, each nanorobot can sense the environment and its neighbours within the perception range in all directions. The fitness calculation of nanorobot i can be expressed as

$$F(\mathbf{x}_i(t)) = \frac{\sum_{\{j:G_j(\mathbf{x}_i(t))>0\}} G_j(\mathbf{x}_i(t)) - D_j(\mathbf{x}_i(t))}{N_{\text{landscape}}}, \tag{2}$$

where $j = \{1, 2, \ldots, N_{\text{landscape}}\}$, $G(\mathbf{x}_i(t))$ denotes the observation result of the search space within the perception range (e.g. heights of the landscape) perceived by nanorobot i, $D(\mathbf{x}_i(t))$ is the Euclidean distance between perceived landscape point and the nanorobot position, and $N_{\text{landscape}}$ is the number of landscape points perceived at the nanorobot position. The resulting fitness is the average height of all landscape points that the nanorobot perceives. As a result, the

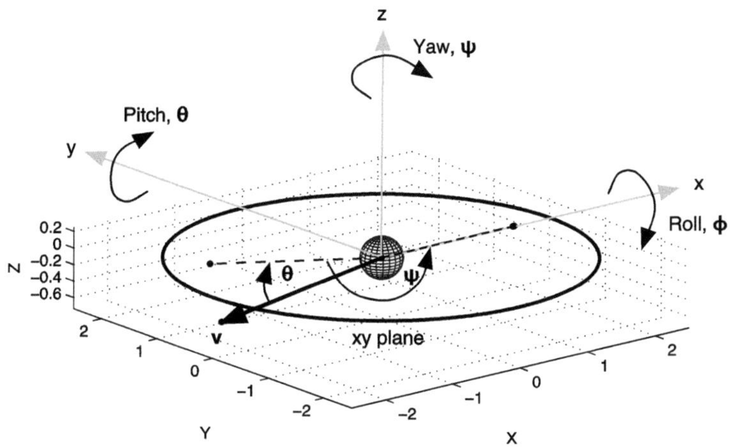

Fig. 3. Euler angles representing the rotation angles of vectors to determine local axes according to nanorobot velocity; note that only pitch and yaw angles are used as it is assumed that nanorobots do not tilt (or roll) while travelling

fitness value of the nanorobot that perceives landscape points from far distance can be negative values. Therefore, when the nanorobot perceives no landscape, the fitness value is set to be a large negative value—such as the negative value of nanorobot perception radius or $-R_p$—rather than zero. Each nanorobot uses fitness value to keep track of its personal best position, $\mathbf{x}_{pbest,i}$, as

$$\texttt{if } F(\mathbf{x}_i(t)) \geq F(\mathbf{x}_{pbest,i}) \texttt{ then } \mathbf{x}_{pbest,i} = \mathbf{x}_i(t). \tag{3}$$

For the determination of the local best position to influence nanorobot movement, nanorobots randomly choose neighbouring nanorobots, which will influence the nanorobot to move toward, with probability p. As the performance of each nanorobot in the neighbourhood is unknown to each other, each neighbour can be in either a better or worse position than its current position. The nanorobot generates a random number between 0 and 1, $r \sim U(0,1)$, for each neighbour; the neighbour with $r \geq p$ is chosen. The position of the chosen neighbour will be used as the local best position, $\mathbf{x}_{lbest,i}$. If there is more than one neighbour chosen, $\mathbf{x}_{lbest,i}$ is the average position among selected neighbours as

$$\mathbf{x}_{lbest,i} = \frac{\sum_{\{a:r_a \geq p\}} \mathbf{x}_a(t) + s_a}{N_{\text{neighbour}}}, \tag{4}$$

where s is a random number between $-qR_p$ and qR_p which can be set as a q proportion of the perception radius R_p as s is used to simulate possible noise in sensing units, and $N_{\text{neighbour}}$ is the number of selected neighbours. Not all neighbours are included in the calculation of the neighbourhood observation; among all neighbours, only some randomly selected neighbour a with a random number r_a greater than or equal to the reference probability $p = 0.5$ has influences on the nanorobot i. When there is no neighbour, $\mathbf{x}_{lbest,i}$ is equal to $\mathbf{x}_i$.

The movement of the swarm is governed by interaction rules. Apart from rules to modify velocity of nanorobots according to PPSO, interaction rules involving the connections among nanorobots are added. The details of rules employed in the swarm system are described in the following:

Attraction Effect: Apart from attraction using bioluminescence for a rather long-distance communication in nature, attraction applies in a closer range as well; dispersion-repulsion forces and electrostatic forces are examples. As the attraction effect among entities can be seen in nature, the same can be applied in future nanorobots. Each nanorobot that has already located in an optimal position, for example it locates on the desired surface in a surface coating application, can generate an attraction signal to attract others that are close enough to detect the signal.

At each iteration, each nanorobot perceives attraction signal from existing optimal nanorobots within its perception range. The attraction signal is sensed as the form of signal intensity. The farther the nanorobot is away from an optimal nanorobot, the lower the intensity it perceives. In other words, the intensity of attraction signal gradually decreases as the distance from the signal source increases. Assume that a nanorobot is at the same position as an optimal one, the distance between these two nanorobots is equal to zero and the intensity of attraction signal becomes 100 percent. When these nanorobots are apart for more than the distance $R_{attraction}$, which is the range that an optimal nanorobot can send signal, the intensity becomes zero. Therefore, at the distance of D from an optimal nanorobot, it can perceive the attraction signal with intensity, $I_{attraction} = (1 - \frac{D}{R_{attraction}})100$ percent. When $I_{attraction}$ is greater than zero, the nanorobot will try to move toward the source. If it is very close to an optimal nanorobot (the distance between two nanorobots is less than the attraction cut-off range, $r_{attraction}$) or the attraction signal is greater than the reference intensity, $I_{ref} = (1 - \frac{r_{attraction}}{R_{attraction}})100$ percent, the attraction effect is activated.

When the nanorobot is attracted to an optimal nanorobot on the desired surface and the attraction effect is activated, the attraction force applies and pull the nanorobot to the nearest available connection of the optimal nanorobot whose attraction force applied on the nanorobot is strongest; at the same time, the repulsion force applies to prevent collision and, consequently, it is located with a small distance from the optimal one. Nevertheless, attraction rules must be set specifically for different tasks. The conditions for attraction rules are often the intensity of attraction signal the nanorobot detecting at its position. The rules decide whether the nanorobot will be attracted and finally connected to an optimal nanorobot. In Fig. 4, examples of the rules regarding the attraction effect for each nanorobot are illustrated. For a simple surface coating problem with one type of nanorobots, no additional attraction rules is required; the nanorobot simply connects to the nearest available connection of the nearest optimal nanorobot. For a surface coating problem with two types of nanorobots creating a checker-board pattern, each type of nanorobots may have its distinctive attraction signal, and the attraction rules can be shown in Fig. 4.

Velocity Update: The velocity of each nanorobot according to PPSO is

$$\mathbf{v}_i(t+1) = w.\mathbf{v}_i(t) + c_1 r_1 \Big(\mathbf{x}_{pbest,i} - \mathbf{x}_i(t)\Big) + c_2 r_2 \Big(\mathbf{x}_{lbest,i} - \mathbf{x}_i(t)\Big) \quad (5)$$

where $\mathbf{v}_i(t)$ is the velocity of nanorobot i at time t, $\mathbf{x}_i(t)$ is the current position of the nanorobot, $\mathbf{x}_{pbest,i}$ is the personal best position of nanorobot i, $\mathbf{x}_{lbest,i}$ is the local best position of the nanorobot, c is the acceleration constant, and r is a random number between -1 and 1. The acceleration constants c_1 and c_2 controls the impact of the personal experience and the social knowledge on the new position of nanorobot. When there is no neighbouring nanorobot nearby, only the previous velocity and the personal best position affect the new velocity.

In PPSO, both the personal best position and local best position are included in the rule to update velocity for the next iteration. The personal best position is the position the nanorobot has visited that resulted with highest fitness values. In real physical applications, the information of nanorobot position with respect to world axes may not be possible. For NARSS, the personal best position is calculated by taking the summation of changing velocity (with respect to its current local axes) on which the nanorobot travels after being on the previous position with highest fitness values. If the nanorobot travels as controlled without any difficulties—for example, noise, friction and collision—then the calculation of the personal best position is accurate. Though some of these difficulties are often present in real applications, it is the only way to refer the personal best position in later iterations given the limited capabilities in the nanorobots. Hence, the calculation of the personal best position for the velocity update is expressed as

$$D_{pbest}(\mathbf{x}_i(t)) = \mathbf{x}_{pbest,i} - \mathbf{x}_i(t) \approx \sum_{step=T_{pbest}}^{t} \mathbf{v}_i(step), \quad (6)$$

where $D_{pbest}(\mathbf{x}_i(t))$ is the distance of the current position of nanorobot i from the personal best position and T_{pbest} is the iteration that the current personal best position becomes active. For the local best position, in real application the position of neighbours is obtained from sensing units; the incoming data may be in terms of intensity of light or chemical concentration which can be used to calculate the distance from the neighbour to the nanorobot. The calculation of the local best position for the velocity update is expressed as

$$D_{lbest}(\mathbf{x}_i(t)) = \mathbf{x}_{lbest,i} - \mathbf{x}_i(t) \approx \frac{\sum_{\{a:r_a \geq p\}} D_a(\mathbf{x}_i(t)) + s_a}{N_{\text{neighbour}}}, \quad (7)$$

where $D_{lbest}(\mathbf{x}_i(t))$ is the distance of the current position of nanorobot i from the local best position, and $D_a(\mathbf{x}_i(t))$ is the distance of the neighbour a from the current nanorobot. As the attraction effect is introduced, the attraction source can be calculated similarly to the local best position as

$$D_{attract}(\mathbf{x}_i(t)) = \mathbf{x}_{attract,i} - \mathbf{x}_i(t) \approx \frac{\sum_{b=1}^{N_{\text{attract}}} D_b(\mathbf{x}_i(t))}{N_{\text{attract}}}, \quad (8)$$

where $D_{attract}(\mathbf{x}_i(t))$ is the distance of the current position of nanorobot i from the attraction source (or optimal nanorobot) within the perception, $D_b(\mathbf{x}_i(t))$ is the distance of the optimal nanorobot b from the current nanorobot, and N_{attract} is the number of attraction that the nanorobot perceives. Therefore, the calculation of new velocity in NARSS is modified as

$$\begin{aligned}\mathbf{v}_i(t+1) &= w.\mathbf{v}_i(t) + c_1 r_1 \Big(\mathbf{x}_{pbest,i} - \mathbf{x}_i(t)\Big) + c_2 r_2 \Big(\mathbf{x}_{lbest,i} - \mathbf{x}_i(t)\Big) \\ &\quad + c_3 r_3 \Big(\mathbf{x}_{attract} - \mathbf{x}_i(t)\Big) \\ &= w.\mathbf{v}_i(t) + c_1 r_1 D_{pbest}(\mathbf{x}_i(t)) + c_2 r_2 D_{lbest}(\mathbf{x}_i(t)) \\ &\quad + c_3 r_3 D_{attract}(\mathbf{x}_i(t))\end{aligned} \tag{9}$$

When the attraction is sensed, the impact of attraction between two nanorobots is considered to be much stronger than other influences; hence, c_1 and c_2 are set to zero. The rules regarding the velocity update for each nanorobot are illustrated in Fig. 4. The velocity of each nanorobot is limited to V_{max} to simulate the maximum velocity of which nanorobots are capable in the environment.

Bounce Effect: When a constructed structure is damaged, free nanorobots must be able to find the damage part and repair it by moving toward that area and connecting to remaining optimal nanorobots in the structure. To achieve this, free nanorobots are required to be in the system with the structure and fly around the search space. It is possible that after a while the swarm might move within a small area. To increase the chance that free nanorobots find the damaged area in a static environment, a set of rules called *bounce effect* is introduced. These rules are to change the velocity of a nanorobot randomly after holding the same personal best position for a defined number of iterations, B_{cutoff}. This is inspired by tumbling behaviours that allow bacteria to move in a random direction when the decreasing nutrient level is detected. The rules regarding the bounce effect are shown in Fig. 4.

Similar systems to the design of NARSS exist in the microworld; wasps and termites constructing their nest with complex design are for example. In social insects, nests are made of biological materials in their environment (such as, plant fibres and local soil) chewed and combined with their saliva. The composites are deposited according to common underlying rules in a swarm to build a nest. Nevertheless, the form of construction that utilises the entities of the system as in NARSS is seen in nature as well; workers of weaver ants, *Oecophylla*, can form a bridge between two leafs using their own bodies, so that other individuals can cross the gaps and pull leaf edges to form a nest [47]. When the leafs are held in place, the ants then connect the edges with thread of silk from a mature larvae [47]. Nonetheless, nanorobots in NARSS remain in place in the self-assembled structures as the system is designed to use minimal requirement to achieve their tasks.

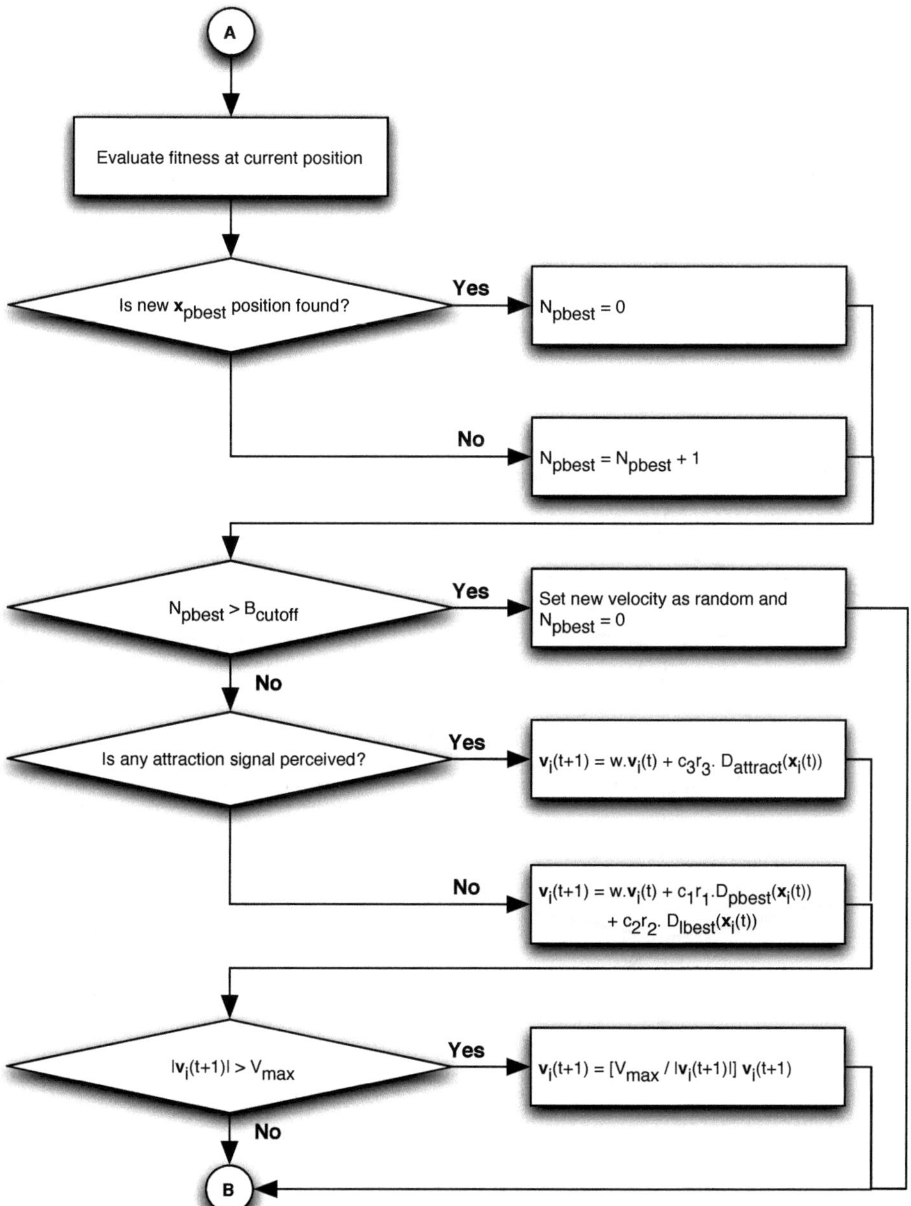

Fig. 4. An example of the rule set for creating a checker-board pattern on the surface

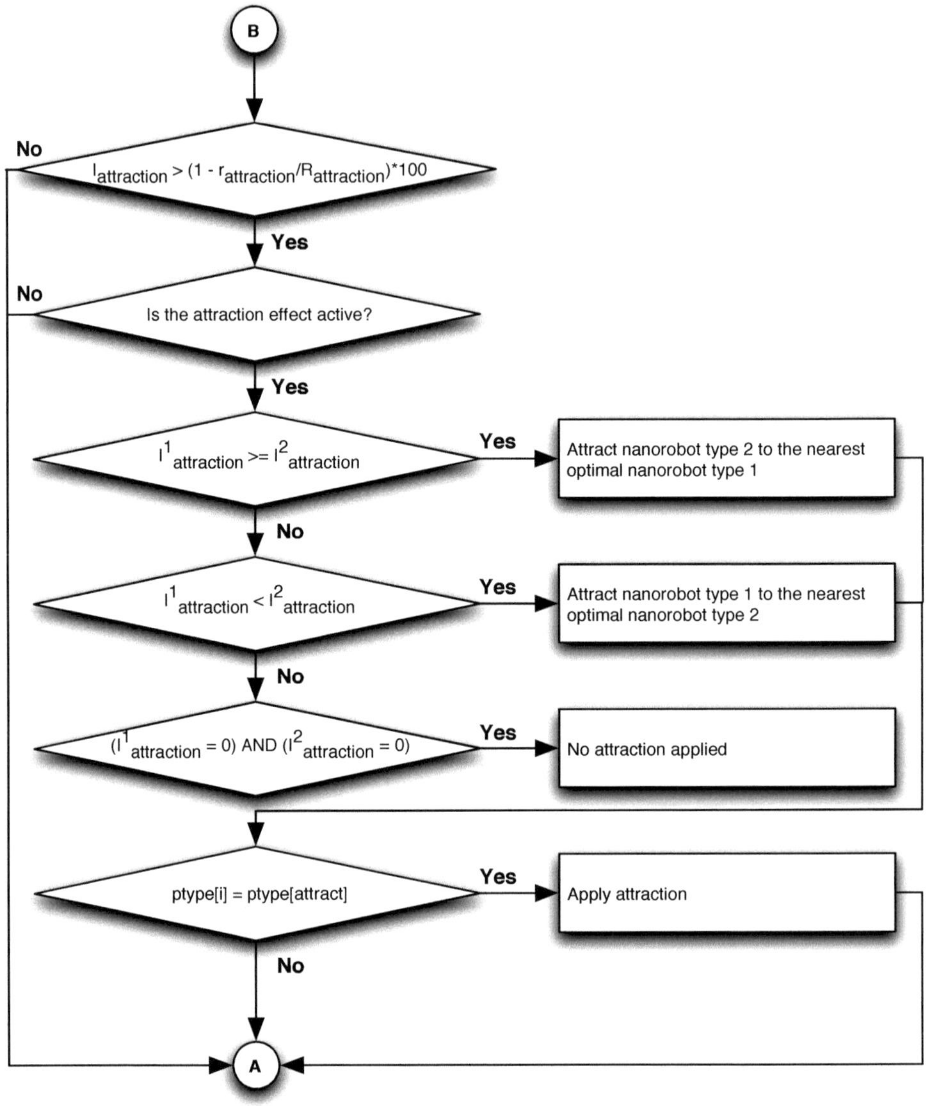

Fig. 4. An example of the rule set for creating a checker-board pattern on the surface (cont.)

However, for effective self-assembly in manufacturing systems, a mechanism to ensure that nanorobots correctly determine the surface as optimal locations when they find the top surface of the platform would be required. For example, the desired surface can be pre-coated with some distinct chemical substance known to nanorobots. When the receptors located on a nanorobot attach to this substance on the desired surface, the nanorobot determines that it is in an optimal position. For the case where more than one nanorobot can reach and stay on the desired surface, they start self-assembly by attracting more nanorobots to them. It can be expected that the orientations of these optimal nanorobots on the surface are various; consequently, further self-assembly from these optimal nanorobots will create fault lines as there is still space on the surface but not enough to fill any nanorobot with acceptable conditions for connection. For many self-assembly applications, such imperfection in the assembly is unwanted. The system will required a mechanism that notifies all free nanorobots when the first optimal nanorobot attaches to the desired surface so that the attraction from existing optimal nanorobots would take precedence rather than the substance on the desired surface. This could be done by broadcasting a distinct signal from the first optimal nanorobots; when a free nanorobot senses the signal, it transmits the signal as well in order to relay the notification signal throughout the swarm. Alternatively, the chemical substance coated on the desired surface could be deactivated (i.e. by changing a chemical property of the substance so that the nanorobot receptor can no longer sense) by the first optimal nanorobot. An example can be found in reproductive systems. When a sperm finds an ovum, the sperm receptors bind to specie-specific ligand in the *zona pellucida* (a tough membrane surrounding the ovum); after penetration, the release of calcium in the egg and cortical granules leads to the inactivation of ligands for sperm receptors to prevent the penetration of multiple sperms to the ovum [49]. In this study, it is assumed that both mechanisms are available to nanorobots; note that no additional characteristic is required in nanorobots to enable these mechanisms.

4 Surface Coating Using Nanorobot Swarm System

One approach to construct molecular machines is to combine simpler mechanisms in surface coating. Surface coating is one of the applications in nanotechnology that is actively researched in industry [1]. Surface coating techniques are used for several purposes including to improve water purification by filtration systems (to detect contaminants on molecular level), to prevent corrosion in substances and to provide stain resistant ability in fabric. In biomedical implants and devices, surface coatings are crucial as coating the surface with bioactive molecules such as collagen can modify physical, chemical and biological properties of devices so that they become biocompatible and inflammatory reactions to the devices are prevented [50]. Self-assembly using future nanorobots is potentially similar to collective constructions in swarms of social insects such as ants and termites. In this study, each nanorobot serves as a building block in the construction of nanostructures. To demonstrate the performance of NARSS, the system is employed in a simple surface coating task, called as *single-layered surface coating*, which is fundamental to building complex structures. The purpose is to cover a desired surface area with a layer of nanorobots. This chapter presents some demonstrations of surface coatings; more information can be found in [51].

4.1 System Settings and Validation

In this pilot study, NARSS operates in a closed vacuum cubic with dimension of 40x40x40. For x and y dimensions, it is set as $[-20, 20]$. For z dimension, it is set to $[0, 40]$. The desired surface is a the top surface of the levelled square platform located at the centre of the cubic. The size of the platform is 10 units in height. With the desired surface area of 3x3, the landscape function is

$$landscape(\mathbf{x}) = \begin{cases} 10 & \text{if } -1.5 \leq \mathbf{x} \leq 1.5, \\ 0 & \text{if otherwise.} \end{cases}$$

In this surface coating task, nanorobots are required to find the desired surface and place themselves on top of the surface. Such problem is similar to a two-dimensional function optimisation where the function is $landscape(\mathbf{x})$. NARSS for two-dimensional problem, which operates in three-dimensional space, can be adapted for the task. The additional dimension is physically treated as the height of the nanorobot position. When the position of nanorobot is updated, the nanorobot moves in three dimensions as well.

In this simple coating, there is only one type of nanorobots involves. The size of each nanorbot is set to 0.5 units in diameter and each nanorobot has the perception radius, R_p, of 2.5 units. If the diameter of each nanorobot in the physical world is assumed to be 100 nanometres, the size of the closed environment is 8 micrometres in each dimension. The size of experimental environment may seem too small for real applications; this setting is, however, compromised by the great computation time required to complete the experiments in the simulation.

In NARSS, there are two important factors affecting the success of the assembly tasks: the number of nanorobots in the system and the perception range of each nanorobot. These two factors indicate the *perception coverage area* of the swarm in the environment shown in Fig. 5. The calculation of the perception coverage area (PCA) can be expressed as $\frac{4}{3}\pi R_p^3 N$, where N is the number of nanorobots in the swarm and each nanorobot can perceive in all directions for the perception radius, R_p. The resulting area is the largest possible area that all nanorobots can perceive. Some part of the perception area in one nanorobot may intersect with that of another nanorobot. Nevertheless, the largest perception coverage area is of interest. Let the *perception coverage ratio* (PCR) indicates the comparison between the perception coverage area and the volume of free space in the environment. The perception coverage ratio is expressed as

$$\text{PCR} = \frac{\text{PCA}}{\text{Free space}} = \frac{\frac{4}{3}\pi R_p^3 N}{(40 * 40 * 40) - (10 * 3 * 3)}. \tag{10}$$

As a larger desired surface area—surface of the platform in Fig. 5 for example—requires a larger number of nanorobots to cover, the estimated number of nanorobots required to cover surface area can be calculated. Let us consider a nanorobot as an atom. As an atom connects to another atom with a strong covalent bond as shown in Fig. 6-a. Let us treat the connected nanorobots as in Fig. 6-b where the outer circles or the radius of connection R represent the range

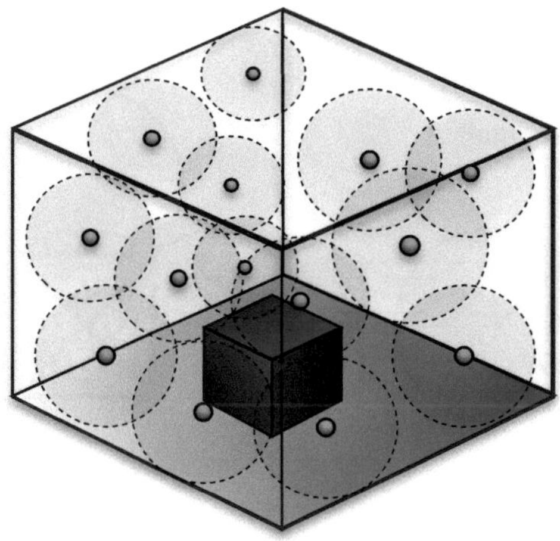

Fig. 5. An example for the perception coverage area in the environment

of attraction and repulsion forces between two connected nanorobots. Let θ be the angle that the attraction and repulsion forces expand for each connection on a plane; in the case where there are m connections that each nanorobot can attach to other individuals on the plane, the angle of each connection becomes $2\pi/m$. After connection, the distance between two nanorobots represented by H is

$$H = 2R\sin(\beta) = 2R\sin\left(\frac{\pi-\theta}{2}\right) = 2R\sin\left(\frac{\pi-\frac{2\pi}{m}}{2}\right), \tag{11}$$

and the intersection range of attraction and repulsion forces between two nanorobots represented by A is

$$A = 2R\cos(\beta) = 2R\cos\left(\frac{\pi-\theta}{2}\right) = 2R\cos\left(\frac{\pi-\frac{2\pi}{m}}{2}\right). \tag{12}$$

Fig. 6-c shows a nanorobot connected to four other nanorobots. In this case ($m = 4$), the least area covered by this nanorobot is equal to the total area of four triangles. The area of a triangle is

$$\triangle = \frac{1}{2}\left(\frac{H}{2}\right)A = \frac{1}{2}R^2\sin\left(\pi-\frac{2\pi}{m}\right). \tag{13}$$

Thus, the least number of nanorobots required to cover any surface area is

$$N_{\text{required}} = \frac{\text{Surface Area}}{m\triangle} = \frac{2 * \text{Surface Area}}{mR^2\sin\left(\pi-\frac{2\pi}{m}\right)}. \tag{14}$$

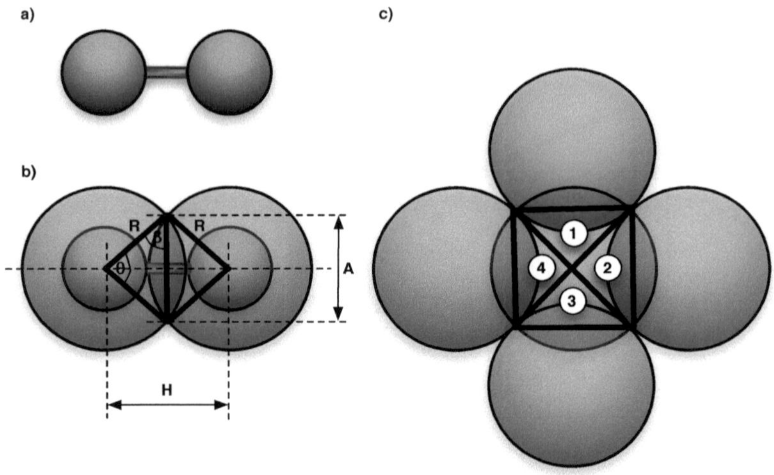

Fig. 6. The connections between nanorobots and the area they cover

Through modelling, the purpose of this study is to investigate the minimal characteristics that are required for future nanorobots. To validate the performance of nanorobot control using the applied swarm intelligence technique, the PPSO algorithm, a series of experiments with varying perception coverage ratio on self-assembly and self-repair tasks. Each is repeated in order to obtain reliable results. The swarm of nanorobots is initially distributed to random positions in the search space. The nanorobots operate as controlled by PPSO and the added interaction rules until reaching any of the following termination criteria:

1. the number of iteration reaches 5,000 iterations, or
2. every nanorobot has movement less than 1.0 units in one iteration.

In order to determine the performance of nanorobots on assembly tasks, two main factors including 1) the number of nanorobots covering on the desired surface to indicate the accuracy of self-assembly and 2) the number of iterations each optimal nanorobot used to indicate the speed that nanorobots can assemble into desired structures. Three experiments are conducted to investigate the effect of changing perception coverage ratio on the performance of the swarm system under different circumstances.

4.2 Experiment 1: Number of Nanorobots versus Perception Range

As the perception coverage ratio is dependent on both the number of nanorobots in the swarm and the perception range of each nanorobot, this experiment investigates the differences of effects from each component to the system performance. For the desired surface area of 3x3, at least 72 nanorobots (estimated by using Equation 14) are required to cover the surface. Fig. 7 demonstrates examples of the coating results at the end of the experiment. The trends of the median result values, called in this study as *perception coverage curve*, for the case of varying number of nanorobots and varying perception range are illustrated in Fig. 9. For the speed of self-assembly, Fig. 8 illustrates time-to-optimal curves for each perception coverage ratio when the number of nanorobots is varied; optimal nanorobots can quickly find the desired surface in the first 500 iterations and the rate becomes slower after 500 iterations. These resulting trends exhibits the exponential growth of $f(x)$, $f(x) = ae^{bx}$; the coefficient a indicates the initial number of iterations of the trend, and the coefficient b indicates the growth constant [52]. The greater the coefficient a, the greater the number of iterations the swarm requires to begin self-assembling. The greater coefficient b indicates the greater rate of the number of iterations between two optimal nanorobots grows exponentially as the self-assembly proceeds. Nevertheless, the initial number of iterations can be distorted from the actual trend in curve fitting process as it is compensated for a better fit; hence, the growth rate reveals a more prominent indicator of self-assembly speed.

Both the accuracy and speed of NARSS on self-assembly tasks increase with the increasing perception coverage ratio. From the observation of swarm movement, the swarm is initially distributed around the free space. Under PPSO-based control, nanorobots interact with the environment and their neighbours to find the desired surface. In a wide free space, nanorobots may gather into several groups; however, individuals can relocate to another group when they are influenced by nanorobots with better positions. A nanorobot finds the desired surface and attaches itself to an optimal location on the surface. Its neighbours interact with this optimal nanorobot and are attracted to move toward the surface. Other nanorobots in the group are similarly encouraged, so they self-assemble on the surface with high speed. Occasionally, nanorobots from nearby groups are recruited and find the desired surface. After a while, nanorobots split into explicit groups. Those that are far from the desired surface have a low chance find the surface. Only nanorobots in the group near the desire surface have the potential to assemble at available location, so the speed of self-assembly decreases.

The performance of NARSS varies with the perception coverage of the swarm. In the varying number of nanorobots case, the swarm density varies. In the varying perception range case, the perception range of each nanorobots varies. The experimental results show that the swarm density has more impact on the effectiveness of self-assembly; the greater number of individuals induces higher chance of multiple interactions, which is a key component of self-organisation. It allows more nanorobots to find the desired surface and self-assemble into a layer on the surface when compared with the case where the perception range is varied. Moreover, regarding the realisation of future nanorobots, the perception range is potentially equal to the operation range of their sensing units and the number of nanorobots can possibly varied to optimise the swarm system; hence, the later experiments in this study primarily vary the number of nanorobots for the changing perception coverage ratio.

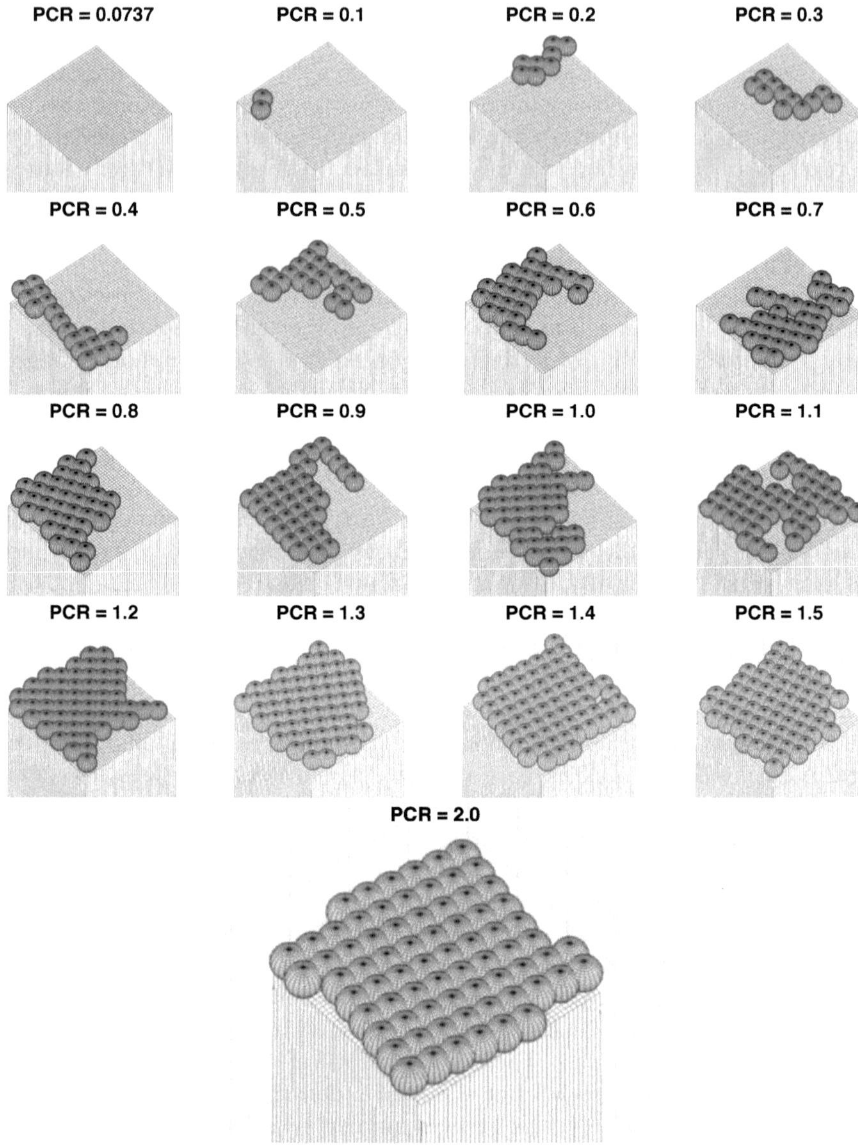

Fig. 7. Optimal nanorobots on the desired surface after 5,000 iterations for each perception coverage ratio in the case where the number of nanorobots is varied

4.3 Experiment 2: Self-assembly and Self-repair

This experiment investigates the ability to cope with a damage to the constructed structure and to repair the damaged structure. NARSS is expected to at least restore the damaged structure to its condition before damaged. This experiment is divided into two parts: *self-assembly* and *self-repair*. The

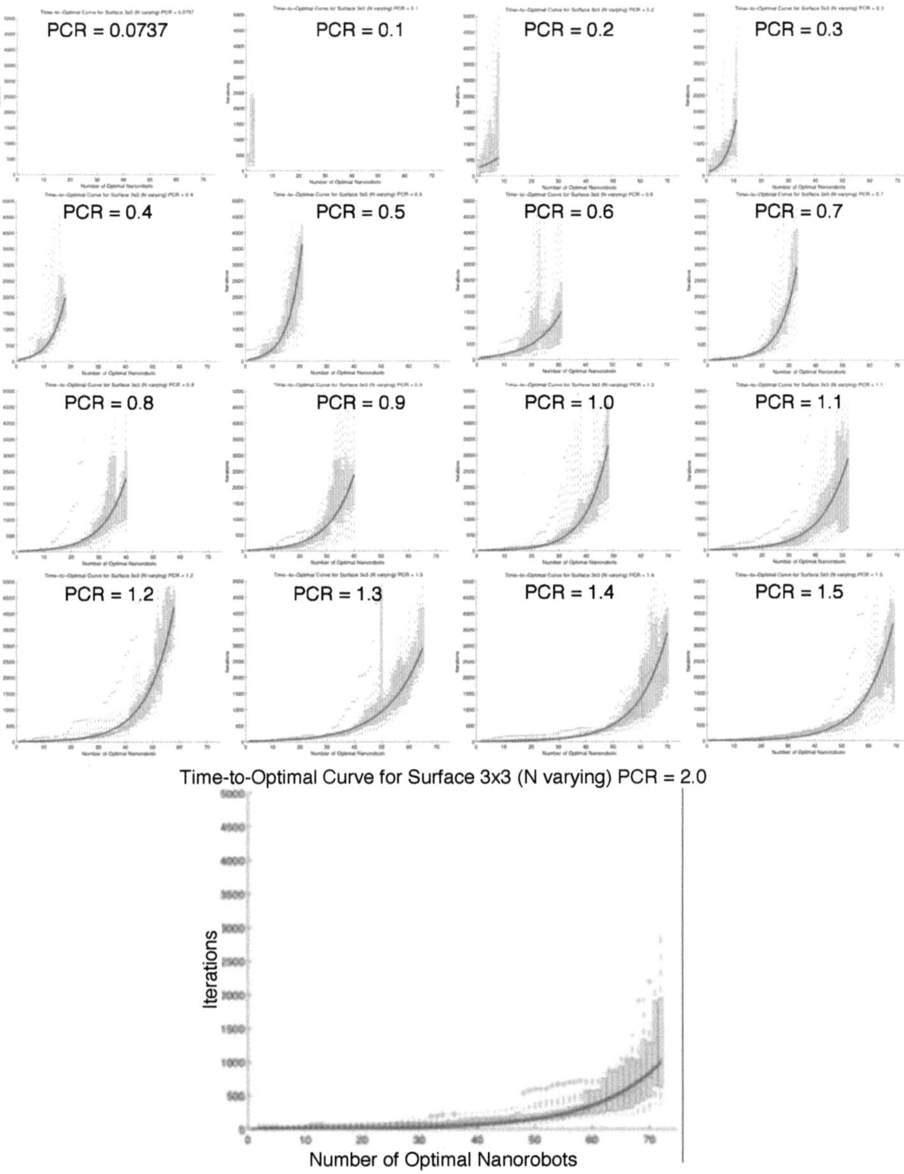

Fig. 8. Time-to-Optimal curves for single-layered surface coating experiment on surface 3x3 when the number of nanorobot, N, is varied; each of which curve shows time-to-optimal curve obtained from specified perception coverage ratio in the relationship between the number of iterations (from 0 to 5000) and the number of optimal nanorobots (from 0 to 75)

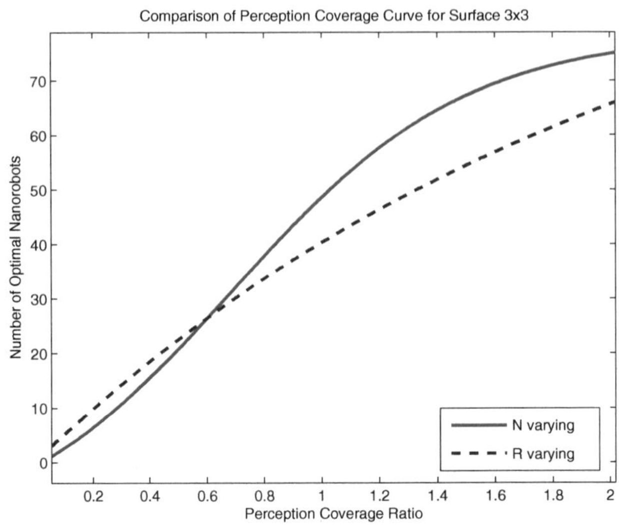

Fig. 9. The comparison of perception coverage curves between the case of varying number of nanorobots (N varying) and varying perception range (R varying)

system begins with self-assembly part which is the same operation as in *Experiment 1* for 2,500 consecutive iterations. After that, the simulation collects all system parameters including the position and velocity of both free nanorobots and optimal nanorobots on the desired surface. These parameters are then used in the self-repair part. The simulation continues the self-assembly part until no nanorobot movement appears or the maximum iterations of 5,000 is reached. In self-repair part, the simulation restores all collected parameters in the self-assembly part at 2,500 iterations. The system simulates a damage to the coated surface by randomly selecting a number of optimal nanorobots and setting their positions and velocities as random. Then, the simulation continues coating for another 2,500 iterations or until no movement occurs. The resulting structure in self-repair part is then compared with the one from the assembly part.

Fig. 10 illustrated the ability to repair the simulated damage. As observed in *Experiment 1*, at 2,500 iterations the groups of nanorobots begin to settle and the self-assembly is mostly done at low speed. A damage to the constructed layer on the surface causes removed nanorobots to become free nanorobots and move around the space again. Each of which has the memory of its personal best position which is one of the available locations on the desired surface. According to the nanorobot control, they move toward these locations and can influence their neighbours to follow. Some of the removed nanorobots and their followers finds the desired surface and repair the damage. The greater the number of optimal nanorobots is removed for a damage, the better chance they can influence other nanorobots and, hence, the faster the self-assembly becomes as shown in Fig. 11. Without a well-defined construction plan, a swarm system of nanorobots can self-assemble into structures through self-organisation. A damage to the

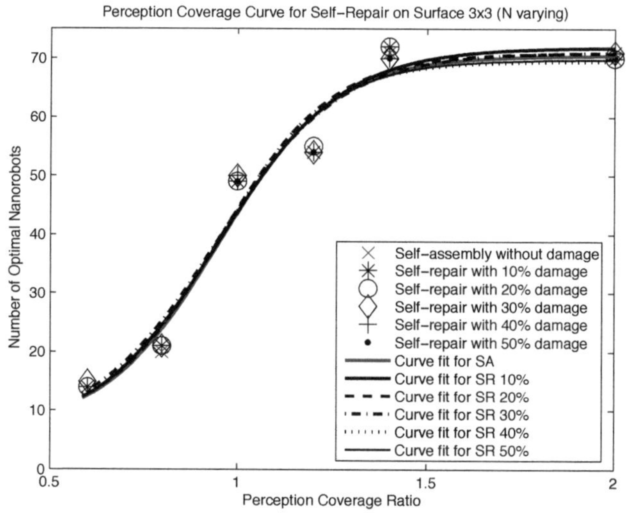

Fig. 10. The comparison of perception coverage curves from self-assembly and self-repair parts in single-layered surface coating experiments on surface 3x3

constructed structure poses as a kind of obstacles for the swarm system to achieve its task. Using self-organisation, NARSS can recover from the damage disturbing the system and repair the structure to its condition before damaged. At different damage rates, NARSS repairs the structure at different speed. The removed optimal nanorobots simulating the damage are placed at random locations in the space; consequently, the swarm diversity is increased. The removed nanorobots induce other nanorobots to move toward the desired surface and speed up the self-repair. For future use of systems of nanorobots, a damage can be introduced to accelerate the process of self-assembly.

4.4 Experiment 3: Noise in Nanorobot Functions

The dynamics in the potential environment around future nanorobots can affect the self-assembly performance of the swarm system. As the principal abilities of the modelled nanorobots in this study are perception and locomotion, the analysis of effects from dynamic environments concentrates on these functions. This experiment investigates the ability of NARSS to cope with noise distorting the nanorobot functions and the impact of distorted functions on self-assembly performance. The dynamic environment effect on the perception of nanorobots can be simulated in NARSS with noise added in the values for nanorobot perception and locomotion. Two noise levels including $k = 1.0$ and $k = 2.0$ (which are 0.4 and 0.8 of the perception radius) are applied to reveal the trend of noise impact on the self-assembly performance.

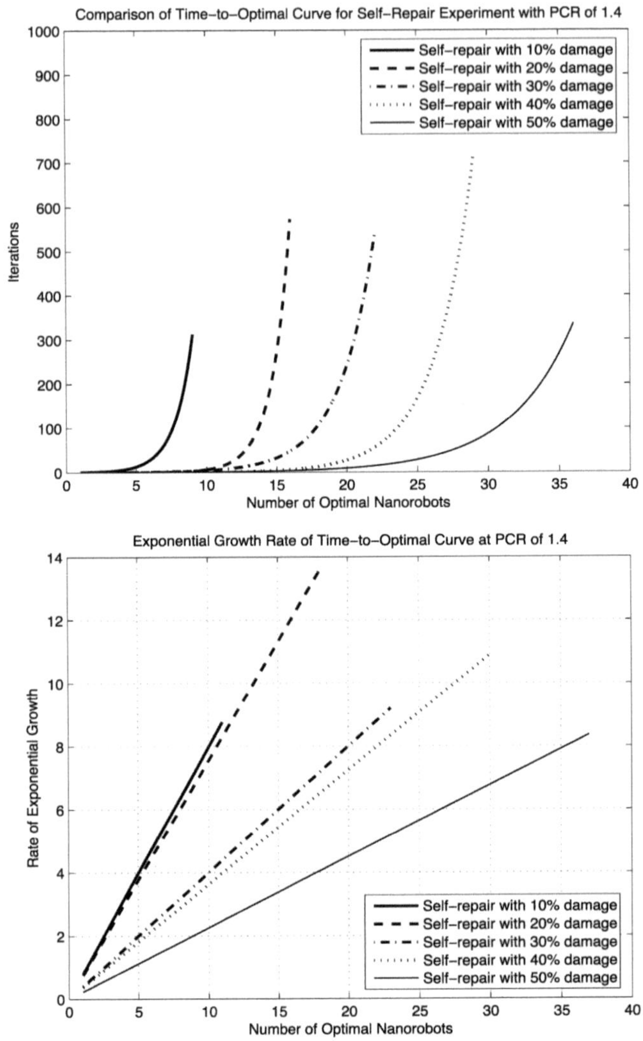

Fig. 11. The comparison of time-to-optimal curves, $f(x) = ae^{bx}$, from self-assembly and self-repair parts on surface 3x3 (PCR = 1.4) where the median number of optimal nanorobots at 2,500 iterations is 68 and their exponential growth rates b; the rate of exponential growth is derived from integrating the rearranged first-order derivative of the exponential growth function: $\frac{df(x)}{dx} = abe^{bx} = bf(x)$, $\frac{df(x)}{f(x)} = bdx$, $\int \frac{df(x)}{f(x)} = \int bdx$, $\ln\left(\frac{f(x)}{f(0)}\right) = bx$, and $\ln\left(\frac{f(x)}{a}\right) = bx$

Compared with the system without noise, NARSS with additional noise self-assembles into a layer on the desired surface with better performance in both the accuracy and speed of self-assembly and self-repair as shown in Fig. 12 and 13. When noise applied in both nanorobot perception and locomotion, experimental results suggest that the effects of both dynamics on the self-assembly performance investigated in terms of the accuracy and speed of self-assembly are integrated. On the one hand, noise added in nanorobot perception affects the controlled movement of nanorobot locally when nanorobots perceive either the landscape or their neighbours. The distorted perception leads the swarm to explore the space more when nanorobots perceive the landscape. As nanorobots have no priori knowledge of other individuals' performance and the neighbours are chosen to influence the next nanorobot movement with the probability of 0.5, noisy perception has little effects on the selection of neighbouring nanorobots. On the other hand, noise in nanorobot locomotion alters the position of nanorobots which affects the controlled movement globally based on inaccurate information. Consequently, noise in nanorobot locomotion brings about random motion in nanorobots and the nanorobot control in this circumstance becomes ineffective; nanorobots can correctly perceive the environment but cannot precisely move according to the nanorobot control. Nevertheless, noise in nanorobot locomotion allows nanorobots to expand their search range. The diversity of the swarm increases and the chance to find the desired surface becomes greater. The system relies on the chance gained from the increased swarm diversity. The increase in swarm diversity can speed up the self-assembly and self-repair even without the influence from removed nanorobots to lead their neighbours toward the desired surface. The composition of noise in both nanorobot functions induces even more swarm diversity than other systems. The increment of swarm exploration bring about more chance to find the desired surface. With a higher noise level, the system is encouraged for more exploration and gains a higher chance to find the desired surface and can, moreover, speed up the self-assembly.

Even though without effective nanorobot control for NARSS in dynamic environments, nanorobots travelling in random motion can successfully self-assemble into a layer on the desired surface with improved performance when compared with those under noise-free circumstances. This is due to the attraction effect applied when free nanorobots move close to the optimal nanorobots existing on the desired surface. The attraction effect between nanorobots is crucial for self-assembly with random motion; otherwise, nanorobots would need to find the desired surface and connect to the existing optimal nanorobots on the surface completely by chance. Though additional noise in nanorobot perception can distort the perception of attraction signal as well, less dramatic outcome on the attraction effect is obtained when the number of optimal nanorobots on the surface becomes greater. It is demonstrated that nanorobots moving with random motion and featuring the attraction effect to attract nanorobots to optimal locations can be adequate for self-assembly tasks; the characteristics of nanorobots identified in section 3.2 indicate minimal features required for future nanorobots, and even simpler nanorobot control than the proposed method using swarm intelligence in section 3.3 is sufficient. Nevertheless, the experiments are conducted in a simple surface coating task. The findings in this chapter are not ensured for

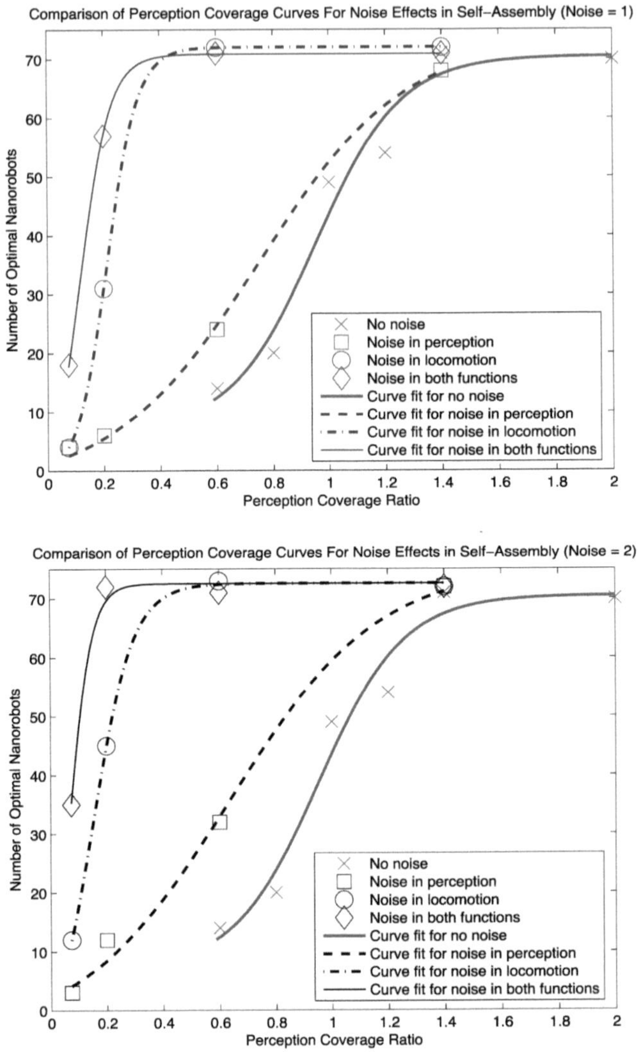

Fig. 12. The comparison of perception coverage curves in self-assembly part with different noise levels applied in both the perception and locomotion of nanorobots

highly complex construction tasks. For the self-assembly of such complex tasks, it is likely that some nanorobot control with some forms of communication among nanorobots would be required.

For more complicated tasks of surface coatings, Fig. 14 demonstrated layering two types of nanorobots on the desired surface and Fig. 15 illustrated patterning nanorobots into checker-board patterns.

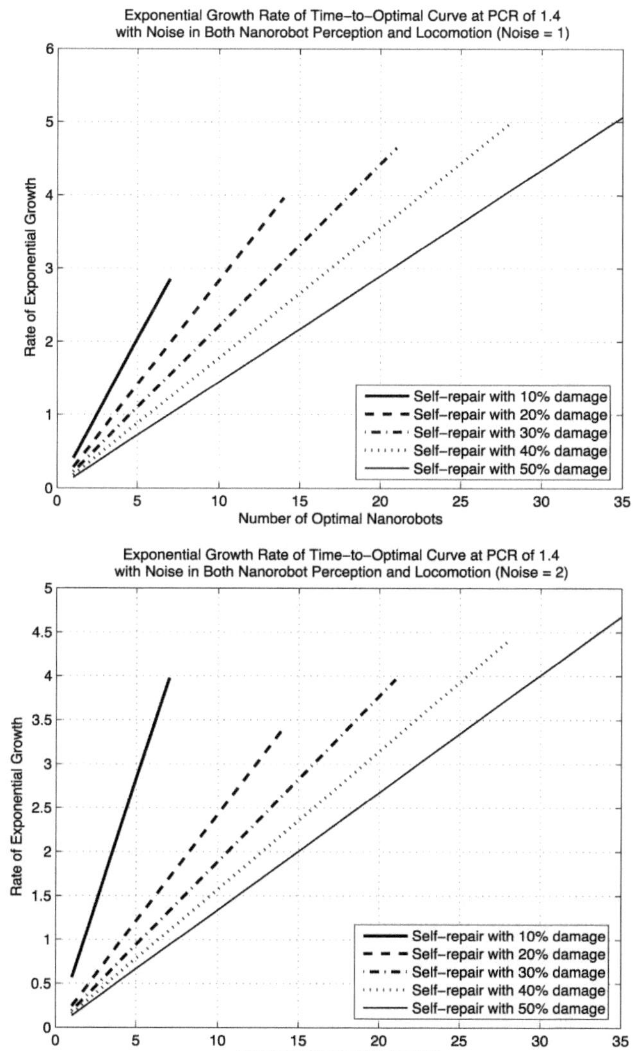

Fig. 13. The exponential growth rates b of time-to-optimal curves, $f(x) = ae^{bx}$, after the damage for the experiment with perception coverage ratio of 1.4 and different noise levels in both nanorobot perception and locomotion; the rate of exponential growth is derived from integrating the rearranged first-order derivative of the exponential growth function. Lines closer to horizontal are better as they indicate slower worsening of time-to-optimal for the increasing number of optimal nanorobots; hence, the speed of self-assembly is faster.

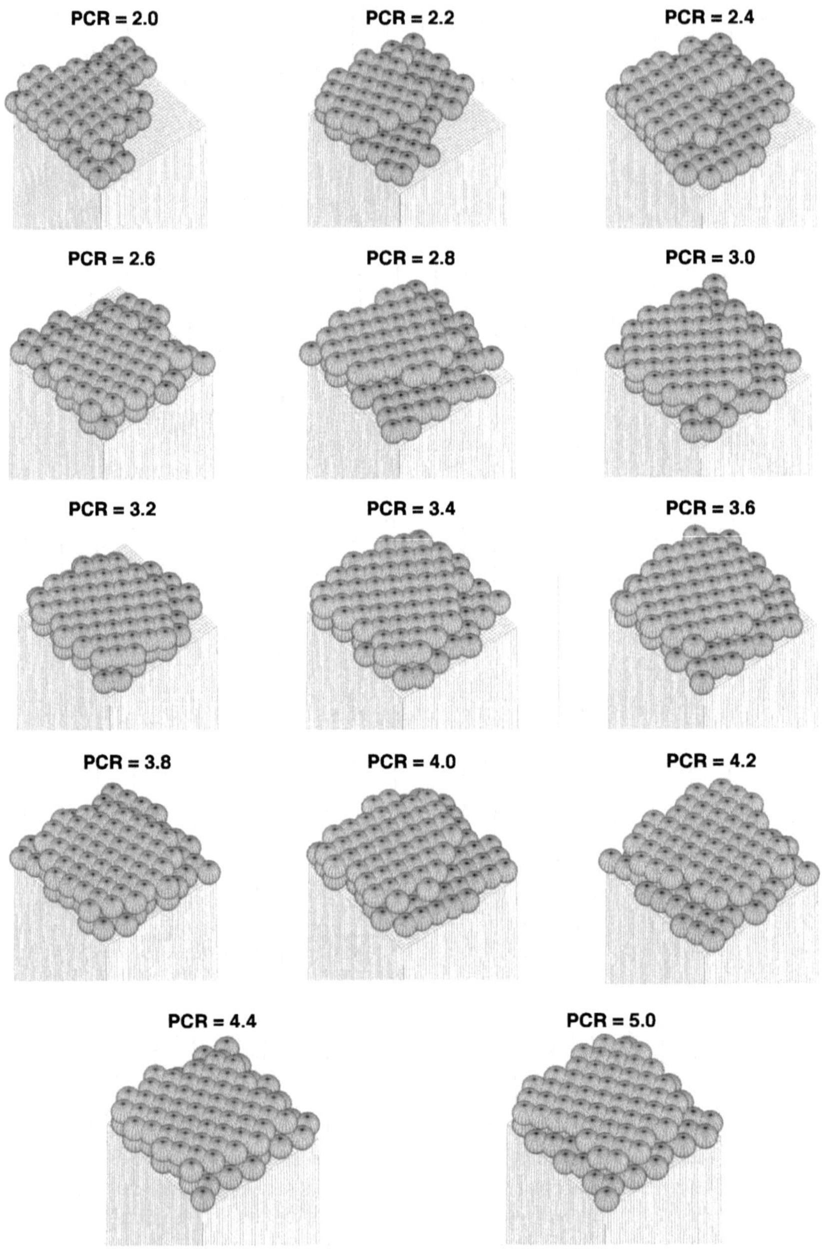

Fig. 14. Optimal nanorobots on the desired surface after 5,000 iterations for each perception coverage ratio; the nanorobots from the first type are shown in black while those from the second type are shown in blue

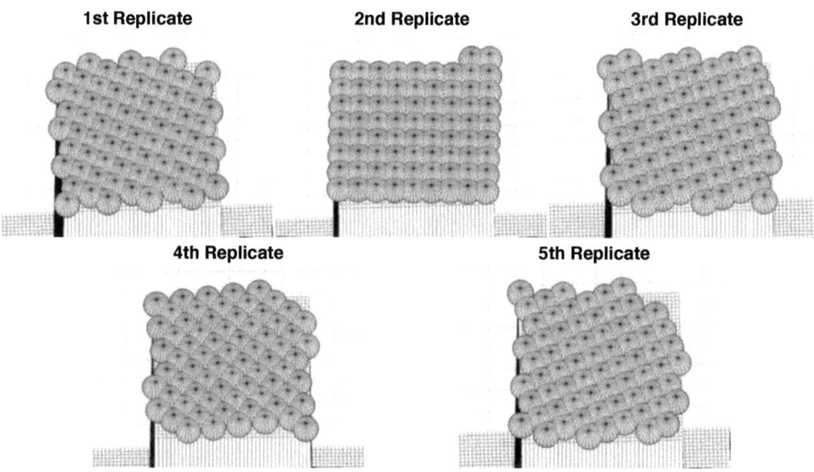

Fig. 15. Optimal nanorobots forming checker-board patterns on the desired surface after 5,000 iterations

5 Future Directions

1. As the greater swarm diversity induces a greater chance to find the optimal locations and self-assemble with higher speed, with the identified characteristics in a static environment nanorobots under PPSO-based control require the greater swarm diversity for self-assembly in spacious environment. *The improvement for a greater swarm diversity in PPSO-based control may allow more effective control for complex self-assembly.*
2. For a complicated self-assembly where effective nanorobot control mechanisms will be needed, additional potential characteristics will be requisite. *The additional potential features to simulate more advanced nanorobots will be added in nanorobot swarm systems.*
3. With a great number of nanorobots, the simulation using sequential programming takes several weeks or months for an experiment. *The simulation of nanorobots processing for self-assembly in parallel will be beneficial.*

6 Conclusion

This chapter has shown that nanorobot control can be modelled using swarm intelligence. It has provided evidence that such nanorobots are capable of producing the emergent functionality of self-assembly and self-repair. Consequently, this study has discovered some significant minimal characteristics and parameters required for the design of useful nanorobots in the future. For a simple self-assembly application, it has been demonstrated in this study that nanorobots with these minimal characteristics can successfully perform self-assembly and self-repair in simulated environment. In [51], more investigation is provided.

References

1. Roco, M.C.: Nanoscale science and engineering: Unifying and transforming tools. American Institute of Chemical Engineers 50, 890–897 (2004)
2. Wolfe, J.: Top 10 nanotech products of 2004. Forbes/Wolfe Nanotech Report 3, 1–3 (2004)
3. Paull, R.: The top ten nanotech products of 2003. Forbes/Wolfe Nanotech Report (2003)
4. Keren, K., Berman, R.S., Buchstab, E., Sivan, U., Braun, E.: DNA-templated carbon nanotube field-effect transistor. Science 302, 1380–1382 (2003)
5. Nguyen, T.D., Tseng, H.R., Celestre, P.C., Flood, A.H., Liu, Y., Stoddart, J.F., Zink, J.I.: A reversible molecular valve. Proceedings of the National Academy of Sciences of the United States of America 102, 10029–10034 (2005)
6. Falvo, M.R., Clary, G.J., Taylor II, R.M., Chi, V., Brooks, F.P., Washburn, S., Superfine, R.: Bending and buckling of carbon nanotubes under large strain. Nature 389, 582–584 (1997)
7. Levit, C., Bryson, S.T., Henze, C.E.: Virtual mechanosynthesis. In: Proceedings of the Fifth Foresight Conference on Molecular Nanotechnology (1997)
8. Kaewkamnerdpong, B., Bentley, P.J.: Computer science for nanotechnology: Needs and opportunities. In: Proceedings of the Fifth International Conference on Intelligent Processing and Manufacturing of Materials (2005)
9. Kaewkamnerdpong, B., Bentley, P.J., Bhalla, N.: Programming nanotechnology: Learning from nature. In: Advance in Computers, vol. 71, pp. 1–37. Elsevier, Amsterdam (2007)
10. Kennedy, J., Eberhart, R.: Particle swarm optimization. In: Proceedings of the IEEE International Conference on Neural Networks, pp. 1942–1948 (1995)
11. Feynman, R.P.: There's plenty of room at the bottom: An invitation to enter a new field of physics. In: Gilbert, H.D. (ed.) Miniaturization, Reinhold (1961)
12. Drexler, K.E.: Engines of Creation: The Coming Era of Nanotechnology. Anchor Press (1986)
13. Drexler, K.E.: Nanosystems: Molecular Machinery, Manufacturing, and Computation. John Wiley & Sons, Chichester (1992)
14. Shirai, Y., Osgood, A.J., Zhao, Y., Kelly, K.F., Tour, J.M.: Directional control in thermally driven single-molecule nanocars. Nano Letters 5, 2330–2334 (2005)
15. Liao, S., Seeman, N.C.: Translation of DNA signals into polymer assembly instructions. Science 306, 2072–2074 (2004)
16. Sherman, W.B., Seeman, N.C.: A precisely controlled DNA biped walking device. Nano Letters 4, 1203–1207 (2004)
17. Seeman, N.C.: From genes to machines: DNA nanomechanical devices. Trends in Biochemical Sciences 30, 119–125 (2005)
18. Cavalcanti, A., Freitas Jr., R.A.: Nanosystem design with dynamic collision detection for autonomous nanorobot motion control using neural networks. In: Proceedings of the International Conference on Computer Graphics and Vision (2002)
19. Cavalcanti, A., Freitas Jr., R.A.: Nanorobotics control design: A collective behavior approach for medicine. IEEE Transactions on NanoBioScience 4, 133–140 (2005)
20. Cavalcanti, A.: Assembly automation with evolutionary nanorobots and sensor-based control applied to nanomedicine. IEEE Transactions on Nanotechnology 2, 82–87 (2003)
21. Holland, O.E., Melhuish, C.R.: Getting the most from the least lessons for the nanoscale from minimal mobile agents. In: Proceedings of the Fifth International Workshop on Aritificial Life (1996)

22. Blackwell, T.M., Bentley, P.J.: Improvised music with swarms. In: Proceedings of the IEEE Congress on Evolutionary Computation, vol. 2, pp. 1462–1467 (2002)
23. Schoonderwoerd, R., Holland, O., Bruten, J.: Ant-like agents for load balancing in telecommunication networks. In: Proceedings of the First International Conference on Autonomous Agents, pp. 209–216 (1997)
24. Ujjin, S., Bentley, P.J.: Particle swarm optimization recommender system. In: Proceedings of the IEEE Swarm Intelligence Symposium, pp. 124–131 (2003)
25. Whitesides, G.M., Grzybowski, B.: Self-assembly at all scales. Science 295, 2418–2421 (2002)
26. Ball, P.: The Self-Made Tapestry: Pattern Formation in Nature. Oxford Press, Oxford (1999)
27. Goodsell, D.S.: Bionanotechnology: Lessons from Nature. Wiley, Chichester (2004)
28. Sendova-Franks, A.B., Franks, N.R.: Self-assembly, self-organization and division of labour. Philosophical Transactions: Biological Sciences 354, 1395–1405 (1999)
29. Flores, H., Lobaton, E., Méndex-Diez, S., Tlupova, S., Cortez, R.: A study of bacterial flagellar bundling. Bulletin of Mathematical Biology 67, 137–168 (2005)
30. Soong, R.K., Bachand, G.D., Neves, H.P., Olkhovets, A.G., Craighead, H.G., Montemagno, C.D.: Powering an inorganic nanodevice with a biomolecular motor. Science 290, 1555–1558 (2000)
31. Daniels, R., Vanderleyden, J., Michiels, J.: Quorum sensing and swarming migration in bacteria. FEMS Microbiology Reviews 28, 261–289 (2004)
32. Anglerfish. Encyclopædia Britannica (2007), `http://search.eb.com/eb/article-9007571`
33. Bioluminescence. Encyclopædia Britannica (2007), `http://search.eb.com/eb/article-8261`
34. Costa-Fernandez, J.M.: Optical sensors based on luminescent quantum dots. Analytical and Bioanalytical Chemistry 384, 37–40 (2006)
35. Kuo, M.D., Waugh, J.M., Elkins, C.J., Wang, D.S.: Translating nanotechnology to vascular disease. In: Greco, R.S., Prinz, F.B., Smith, R.L. (eds.) Nanoscale Technology in Biological Systems. CRC Press, Boca Raton (2005)
36. Wagner, P.: Nanobiotechnology. In: Greco, R.S., Prinz, F.B., Smith, R.L. (eds.) Nanoscale Technology in Biological Systems. CRC Press, Boca Raton (2005)
37. Norton, J.A.: Nanotechnology and cancer. In: Greco, R.S., Prinz, F.B., Smith, R.L. (eds.) Nanoscale Technology in Biological Systems. CRC Press, Boca Raton (2005)
38. Winfree, E.: DNA computing by self-assembly. The Bridge 33, 31–38 (2003)
39. Seeman, N.C.: Biochemistry and structural DNA nanotechnology: An evolving symbiotic relationship. Biochemistry 42, 7259–7269 (2003)
40. Benenson, Y., Paz-Elizur, T., Adar, R., Keinan, E., Livneh, Z., Shapiro, E.: Programmable and autonomous computing machine made of biomolecules. Nature 414, 430–434 (2001)
41. Benenson, Y., Gil, B., Ben-Dor, U., Adar, R., Shapiro, E.: An autonomous molecular computer for logical control of gene expression. Nature 429, 423–429 (2004)
42. Gardner, T.S., Cantor, C.R., Collins, J.J.: Construction of a genetic toggle switch in. Escherichia Coli 403, 339–342 (2000)
43. Gardner, T.S.: Genetic applets: Biological integrated circuits for cellular control. In: IEEE International Solid-State Circuits Conference: Digest of Technical Papers, vol. 44, pp. 112–113 (2001)
44. Payton, D., Daily, M., Estowski, R., Howard, M., Lee, C.: Pheromone robotics. Autonomous Robot. 11, 319–324 (2001)

45. Kaewkamnerdpong, B., Bentley, P.J.: Perceptive particle swarm optimisation. In: Proceedings of the Seventh International Conference on Adaptive and Natural Computing Algorithms, pp. 259–263 (2005)
46. Kaewkamnerdpong, B., Bentley, P.J.: Perceptive particle swarm optimisation: An investigation. In: Proceedings of the IEEE Swarm Intelligence Symposium (2005)
47. Bonabeau, E., Dorigo, M., Theraulaz, G.: Swarm Intelligence from Natural to Artificial Systems. Oxford University Press, Oxford (1999)
48. Stevens, B.L., Lewis, F.L.: Aircraft Control and Simulation, 2nd edn. Wiley Interscience, Hoboken (2003)
49. Speroff, L., Fritz, M.A.: Clinical Gynecologic Endocrinology and Infertility: The Cervical Spine Research Society. Lippincott Williams & Wilkins (2004)
50. Hildebrand, H.F., Blanchemain, N., Mayer, G., Chai, F., Lefebvre, M., Boschin, F.: Surface coatings for biological activation and functionalization of medical devices. Surface & Coatings Technology 200, 6318–6324 (2006)
51. Kaewkamnerdpong, B.: Modelling Nanorobot Control using Swarm Intelligence. PhD thesis, University College London (2009)
52. Weisstein, E.W.: Exponential growth (MathWorld–A Wolfram Web Resource), `http://mathworld.wolfram.com/ExponentialGrowth.html`

11

ACO Hybrid Algorithm for Document Classification System

Nikos Tsimboukakis and George Tambouratzis

Institute for Language and Speech Processing, 6 Artemidos & Epidavrou Str. Amaroussion, Greece
{ntsimb,giorg_t}@ilsp.gr

Abstract. In the present study an ACO algorithm is adopted as a part of a document classification system that classifies documents written in Greek, in thematic categories. The main purpose of the ACO module is to create a word map that will assist in the representation of the documents in the pattern space. The word map creation algorithm proposed involves additional deterministic sub-routines and aims at clustering together into groups thematically-related words. The performance of the proposed system is compared with an alternative system implementation that is based on the established SOM neural network.

1 Introduction

The considerable expansion of the usage of the internet over many different aspects of people's life (as evidenced by blogs, social networks, e-mail etc.) in recent years has highly increased the amount of freely available information. Additionally, every modern business uses electronic database systems to handle its records and files. Information from both these aforementioned sources is mostly expressed in many different varieties of natural language. The organization of linguistic information in a meaningful manner, based on content, is extremely important to support and simplify the retrieval process and to enhance database systems via the use of intelligent query engines. The process of manual topic indexing is feasible in only a few cases and it is generally accepted to be ineffective for extensive collections of data. On the other hand, machine learning systems that are able to adapt and learn from real-world data, can handle the organization task cost-effectively and with consistency. To that end the system presented here aims at classifying documents written in the Greek language to categories with respect to the subject they are dealing with.

The majority of machine learning algorithms require that the patterns processed must be represented as a set of features. This set of features must reflect the patterns' characteristics and is most frequently implemented as a numeric vector. Several document organization systems have been studied, such as the WEBSOM (Kohonen et al. 2000 and Lagus et al. 2004) for example, which is based on the Self-Organising Map (SOM) neural network model. The WEBSOM operates by representing each document with a number of TF-IDF features (Drucker et al. 1999 and Salton et al. 1998). The TF-IDF features are actually word-based and the amount of features

C.P. Lim et al. (Eds.): Innovations in Swarm Intelligence, SCI 248, pp. 215–236.
springerlink.com

available is equal to the total number of words in the whole dataset. It is known that the number of words and in particular words that appear few times in a typical collection is very large and thus a subset of the TF-IDF features is finally selected for the representation (MacKay 2003). Furthermore, often a transformation of the input space to a lower dimensionality is adopted, such as random mapping (Kaski 1998) or singular value decomposition (SVD), in order to reduce the system complexity and speed up the training process. In the present study an alternative document representation is examined which aims at representing documents in a more compact and comprehensive form by adopting a thematic word map. The word map is created via a completely automated process that is based on an Ant Colony Optimization metaheuristic. The effectiveness of the document representation produced is then examined in a document classification task and the proposed system is compared on the basis of its performance with those created by more conventional machine learning systems.

This chapter is organized as follows: In section 2 the motivation and principles of the ACO-based algorithm are presented. In section 3 the proposed architecture for the document classification system is described. The MLP classifier is analyzed in section 4 and in section 5 an alternative implementation using a SOM architecture is provided for comparison purposes. The experimental results for a real world database are detailed in section 6. The final section of this chapter comprises concluding remarks on the experiments performed and suggestions for future work.

2 Principles of the Proposed Solution

Document organization systems adopt a feature selection procedure to represent documents in a numeric format in the feature space, based on the specific features selected. Hence, the effectiveness of each organization system relies heavily on these features. Depending on the organization categories and purposes, or whether they focus on author style or document meaning, a different variety of features is selected (Freeman & Yin 2005). In systems that perform a stylometric analysis, mostly linguistic features are selected such as Part-Of-Speech, Structural and Morphological features (Tweedie et al. 1996). On the other hand, systems that are aimed at performing content analysis in large scale problems (which are not constrained to a small set of key-words) mostly use lemma[1]-based features (Manning & Schütze 1999). These features are based on the frequency of appearance of a given lemma in each document. Several transformations of frequency measurements have been used, such as the well-known TF-IDF measurements and Boolean frequency measurements, over a predetermined threshold. In real-world problems, where document organization systems are focused, the number of lemmas involved can be extremely high and becomes even higher as the document collection size increases further. In order for the systems to be able to cope with this bottleneck, various strategies have been proposed. For example the WEBSOM system uses a fast transformation of the input space into lower dimensions via the use of a randomly-generated stochastic matrix (Kaski 1998) as a replacement to the computationally more demanding Singular Value Decomposition method.

---

[1] Lemma refers here to the word in its basic form without any inflectional suffix (e.g. all three words ‘spoken’, ‘speaks’ and ‘spoke’ would be converted to the form ‘speak’).

To address this issue, the system presented here aims at reducing the dimensionality of a feature space by creating a thematic word map as an intermediate layer. Taking into account that in natural language - apart from the set of common functional words that can appear in every context - there are many words (or lemmas) that are mostly used to describe a specific topic, it would be desirable that words describing similar topics be represented in neighboring areas on the map. This thematic word map is employed in a later phase of the classification process as an index for each document that depicts its content. More specifically, for each document the proportion of its lemmas that belong to a specific map area is counted to form a vector with a dimensionality equal to the number of areas on the map, which contains the frequency of words from each area in the document. Via the thematic map, each document may be represented in a more comprehensive feature space, whose dimension is equal to the number of categories the thematic word map contains.

In order to implement this thematic word map for each lemma, context-based information is extracted and utilized. The basic assumption made is that words which appear in similar contexts in natural language expressions should bear related meanings. The contexts considered here are periods as defined in natural language by the use of major punctuation marks such as full-stops. This basic assumption has also been made in other experiments involving word clustering (for instance Georgakis et al. 2004 and Kohonen & Somervuo 1998). This hypothesis doesn't state that the words within each period should be synonyms or bear an identical meaning, no more than they are used to describe related topics. However, it would be desirable for words that co-appear many times in sentences to be assigned to the same group, as their frequent co-appearance is a strong indication that they are used to express related meanings. By counting word co-appearances within the same period, a strong indication is drawn about their semantic relationship that can be numerically expressed. This numeric value is used as a proximity measure which is input to an Ant Colony Optimization (ACO) algorithm (Kennedy & Eberhart 2001) that aims at placing the nearest (in terms of semantics) words in neighboring groups, or preferably within the same group. At each step of the optimization algorithm, the solution is evaluated by using a function that counts the total similarity among the pairs of words in each sentence as reflected by the thematic map created. Following the generation of several solutions, one for each ant being simulated, the solution that produces the thematic map with the highest similarity measure is finally selected as the feature set for document representation. The ACO thematic word grouping algorithm will be presented in more detail in section 3.1.

3 System Architecture

The system presented in this study comprises three distinct modules.

(1) The first unit implements the preprocessing step where documents are converted to lemmatized form by the use of the ILSP tagger/lemmatizer (Papageorgiou et al. 2000). Since the Greek language possesses a rich morphology, lemmatisation reduces the pattern space and consequently the number of distinct words appearing in documents. Even after these preprocessing actions, the amount of

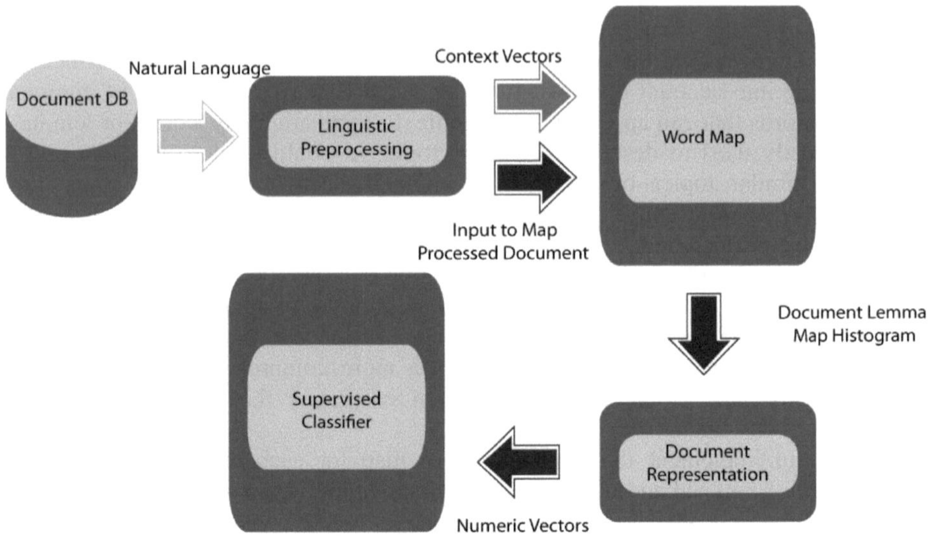

Fig. 1. General architecture of the ACO-based document classification system

different word forms remains quite large (approximately 7,600 lemmas for the dataset used in the experiments detailed in section 6).

(2) The second step comprises the thematic word map module, where words are grouped in terms of their concept using an ACO algorithm. This module serves to create in an unsupervised manner groups of words, where each word group contains words that are related in terms of content. This module employs context-related information accumulated via the sampling of natural language periods.

(3) The final module is the classifier, where a supervised technique is used to assign documents to categories. This is implemented by making use of the groups of words generated from the previous module. Within the present study, a neural network architecture is used to perform the classification, this being an MLP network (Haykin 1999). The classification module was preferred over a clustering module since it allows for a better evaluation procedure, since clustering results can not be compared with a reference clustering in an unambiguous and straightforward manner. Additionally a reference clustering is in most cases difficult to define while the expected classification result can be easily extracted manually from a dataset when the possible categories are predefined. Furthermore, the classification accuracy (the percentage of patterns that is classified correctly in accordance to the preferred scheme) is a simple measure that reflects the degree to which the system has adapted to the requested problem. Since in this piece of research the main concern is the evaluation of the effectiveness of the proposed representation, a classifier module has been considered more appropriate.

3.1 Word Map Module Overview

The implementation chosen for the thematic word map was motivated by several variations of the ACO (for instance Lessing et al. 2004 and Martens et al. 2006). The

ACO metaheuristic is based on replicating physical ant-colonies, where (Dussutour et al. 2004) ants exhibit a collective approach to progressively minimise the path they need to traverse from their nest to the source of food and back, this process being based on laying a chemical substance (pheromone). In analogy, in other optimisation tasks, the artificial ants are independent agents that autonomously search the pattern space for good solutions and then indirectly combine these solutions to find a global-best solution in subsequent epochs.

The ACO was originally applied to the solution of the well-known traveling salesman problem (TSP) (Dorigo & Gambardella 1997 and Stützle & Hoos 2000). In TSP, the aim is to visit all cities from a given set in such an order that the total path length is minimized. In most of the solutions of the TSP problem, for every pair of cities a d_{ij} value is defined which represents the physical distance between city i and city j. Furthermore, in the ACO implementation, for each arc that connects cities i and j a pheromone value τ_{ij} is assigned, which represents the pheromone trail value. Based on the total length of the paths generated by the ants, the pheromone trails are updated and the preferred arcs that contribute to better solutions are gradually assigned stronger pheromone trails. Initially the trail values are randomly assigned and each ant is positioned on a random city from which to initiate its search. At each step of the ACO algorithm, each ant (which is equivalent to a candidate solution) creates a complete TSP solution (route) that connects all cities, by choosing a number of i to j transitions and thus moving from one city to the next one. At each step, the choice of the next city is made based on a probability distribution which assigns probabilities according to a combination of the values d_{ij} and τ_{ij}. Within this selection phase cities that already have been visited are excluded.

In order to map the ACO technique to the thematic word map creation, the proposed implementation considers as the problem states specific words instead of the TSP cities. The main purpose of the ACO is to determine the best solution, in terms of co-appearance similarity that joins all available words within a sentence. The distance of the cities that characterized the original TSP problem has been replaced by co-appearance measurements between pairs of words. Thus, the TSP objective of finding the closest cities is transformed here into the aim of locating the most similar words, in terms of context. Following the simulation of the ACO process, an algorithmic processing of the ACO outcome takes place. During the random selection phase of the ACO where new paths are chosen, word groups are formed. This operation is functionally equivalent to the agglomerative clustering algorithm (Everitt et al. 2001) and involves merging the new word selected at each step with existing groups. These word groups are ordered along a one-dimensional structure according to their similarity, leading to the creation of a data structure resembling an ordered list. The evaluation function that the algorithm tries to improve is based on the total proximity of the words of each sentence. The proximity between two words, according to the grouping created, is defined as the distance between the corresponding groups to which the two words belong.

To summarize, the word map module is based on an ACO-type process, which is repeated for a number of epochs. In each epoch, an 'ant' is simulated which tries to create a good contextual mapping of words to groups. The process for each ant can be divided into 4 steps, as listed below (Fig. 2):

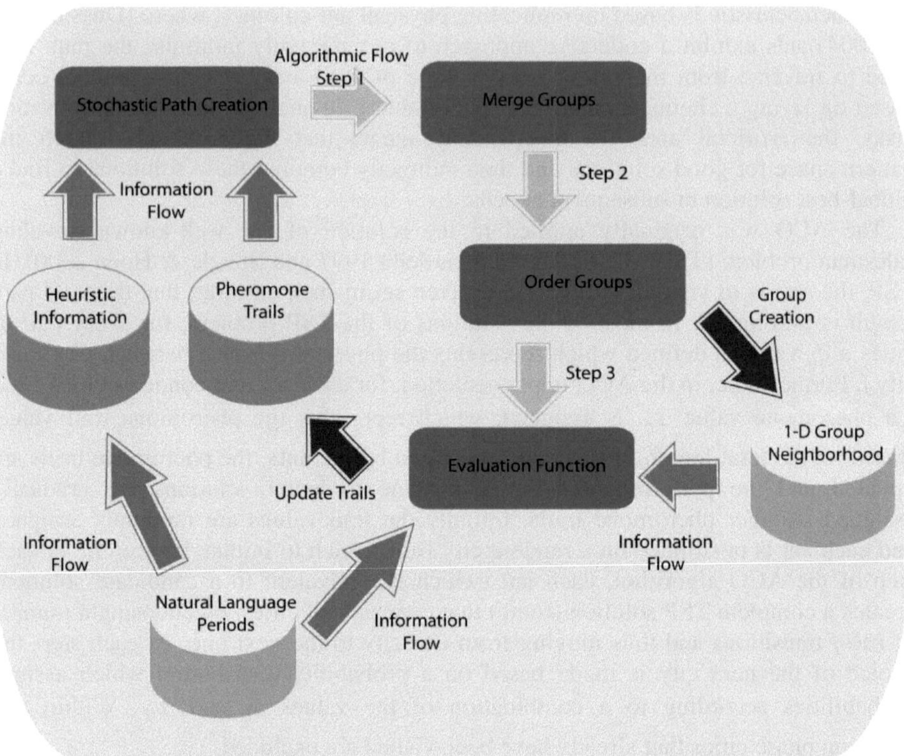

Fig. 2. Overview of the algorithmic flow of the hybrid ACO proposed system

(i) Step 1 – Word clustering: For the current ant, each word is examined in turn and grouped with the one closest to it (to be more accurate, to a word situated close to it in terms of context), based on a stochastic approach taking into account both the pheromone (i.e. previous ant solutions) and the heuristic function. Following this step, a clustering of words is generated.

(ii) Step 2. – Merging: The groups of words generated are then merged, if required, to reach a clustering solution with a given number of groups.

(iii) Step 3. – Linked list creation: The groups are then mapped to a one-dimensional structure, so that most similar groups are assigned to neighboring units.

(iv) Step 4. – Evaluation: The solution of this ant is then evaluated to determine how accurately it maps context.

Following this process, for the given epoch all ants are simulated and the corresponding solutions are then evaluated. Using an iteration-best scheme, the best solution is used to update the pheromone in preparation for the next iteration of the ACO. This process is described in more detail in the following sections.

3.2 ACO – Selection Phase Implementation

Initially from the complete set of documents in natural language all the periods are extracted (each period being defined as the sequence of words between consecutive major punctuation marks). It must be noted that the words in the periods have been converted to lemmatized form, in order to reduce the number of distinct words to be processed. For each pair of words (or lemmas more precisely) the number of times these words co-appear in the same sentence within the entire corpus is counted. This number is divided by the minimum value of the frequencies of appearance of these two words. This measurement represents the similarity between the two words which was denoted earlier as d_{ij}.

$$d_{ij} = \frac{CoApp(i, j)}{\min(Freq(i), Freq(j))} \tag{1}$$

where i and j denote the two words being processed and $CoApp$ corresponds to the frequency of co-appearance of the two words. As defined in eqn. (1), the similarity value of d_{ij} is equal to the value d_{ji} and thus this measurement is symmetric. It can easily be seen that the number of floating-point d_{ij} values required to store in computer memory for all pair of words is proportional to n^2, where n is the number of words (an indicative number for the number of words in the experiments presented later is 7600). On the other hand the square matrix d_{ij} is extremely sparse since only a small fraction of words co-appear even once in the same period throughout the entire corpus. For example, in the dataset used for experimental purposes (see section 6) the sparseness ratio of the square matrix never exceeded 5%, which means that at most 5% of the values are non-zero. In order to overcome the bottleneck of storage space, a sparse matrix implementation was selected, where only non-zero values were stored. The symmetric nature of the measurement adopted further reduces the amount of memory required by half. The implementation selected had an additional index for every word i that comprises all the indexes of the words j that the word i co-appears with (and for which non-zero values need to be stored). This index was utilised for optimization purposes during the ACO algorithm execution.

During the ACO optimization phase, ants are randomly initialized via the following procedure: For each ant initially all available lemmas are randomly arranged. Then each lemma is selected and it is assigned to the same team with another randomly selected lemma according to a probability distribution that encodes both the similarity between the two lemmas and the pheromone trail strength. The following equation expresses the probability p_{ij} that for a given lemma i, lemma j is selected:

$$\mathrm{p}_{ij} = \frac{(d_{ij})^a (\tau_{ij})^b}{\sum_k (d_{ik})^a (\tau_{ik})^b} \tag{2}$$

The τ_{ij} value represents the pheromone strength for the connection between lemmas i and j and it reflects the algorithm tendency to prefer a specific connection (and thus a specific pair of lemmas) over others. Parameters a and b are set to values that allow

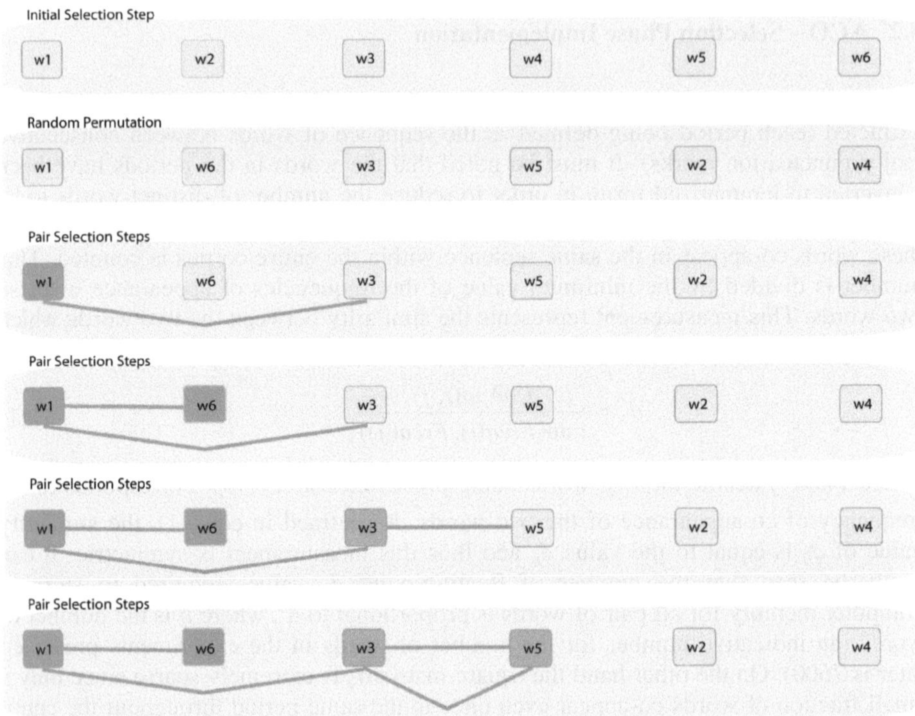

Fig. 3. Selection phase execution example. Following the random permutation step, the following pair selections are made: (i) words w_1 and w_3, (ii) words w_6 and w_1, (iii) words w_3 and w_5 and (iv) words w_5 and w_3. Darker colored words in each selection step indicate words for which an appropriate pair has been selected.

the algorithm to focus more on the heuristic information regarding the suitability of the given solution or on the pheromone trail strength. When the similarity between two lemmas is large enough, the probability distribution expressed by equation (2) enforces the assemblage of these two lemmas to the same group. During this selection phase, if the two lemmas already belong to different groups, then these groups are merged to construct a larger group. The selection procedure proposed here is different in an important way to the one used in the application of the ACO on the TSP. In the TSP case, where an exact solution is provided, the selection phase continues by selecting the most recently visited city to continue, while all the cities that have already been visited are left out. Due to the fact that the number of lemmas in the dataset is very large, the exclusion of the 'visited', so far, lemmas would introduce a possibly excessive degree of complexity at each selection round. To overcome this potential bottleneck, the word permutation step was preferred that is executed at the beginning (as presented in Fig. 3). In addition, the merging method (step 2) is an essential step for combining already grouped lemmas to ensure that the desired number of groups is reached. This modification would not be valid for the TSP problem where the order of visit of each city is very important to the solution created.

For each ant of the population, each selected lemma pair is stored in a matrix to be used later in the algorithm.

3.3 Ordered Group Creation Phase

Following the completion of the selection phase where semantically-similar lemmas are grouped together, it is possible that the lemmas' groups that have been formed are very few in number. In extreme cases there can be only a single group created for a given ant. If this is the case, the corresponding solution needs to be eliminated. To achieve that, it is given a very low evaluation value so that it doesn't have any realistic chance of winning and hence it isn't processed any further. Instead, if the total number of groups is larger than the number of required groups, a merging procedure takes place. It must be noted here that the number of desired groups is a parameter that is provided by the user and which corresponds to the dimensionality of the vector that will be used to represent the documents in the feature space.

Regarding the merging procedure (see step 2 of section 3.1), it is known a priori that the number of further merging actions required is equal to the number of groups already created minus the desired number of groups. Within each action, the two most similar groups are merged to form a new group. The similarities between groups at each step are re-estimated before selecting the most similar groups and performing the new merge. The similarity between two groups is defined in terms of complete linkage (Everitt et al. 2001 and Duda et al. 2001), which is calculated as the average similarity between each word in the first group with each word in the second (the total number of pairs is $m*n$ where m is the size of the first group and n the size of the second). For the similarities between the two groups the d_{ij} values that were presented in equation (1) are used while the pheromone trail strength is ignored at this step.

After the merging of groups, the final number of groups available is equal to the desired number of groups. These groups are positioned on a lattice of nodes according to their similarity with their neighboring nodes to create a topological mapping (Kohonen 1997). This step is implemented as a double-linked list where each node has a right and a left connection. In the case presented here the left and right connections are equivalent and are only used to link a node with its neighbors. Each node in the list is assigned a single word group. Initially the two most similar groups in terms of distance d_{ij} are selected and they are linked so that the left connection of the first group points to the second and the right connection of the second group points to the first. This procedure continues by selecting the next pair of most similar groups until all the groups are placed within the list and the list is fully connected. It must be noted here that circles are not allowed and therefore before a new connection is established it is checked and if it would result in a circular list it is discarded (refer to Fig. 4, step N+2). At the linked list creation step (section 3.1) the similarities (d_{ij}) determined in the previous merging steps are cached in order to make the algorithm implementation more efficient and thus render the simulation faster. This process is of course sub-optimal, as at each step only the locally-optimal pair of groups is examined but has been chosen here to give a sufficiently good result in a limited amount of processing time (since, as shall be described in the experimental section, an exact solution would prove to be computationally intractable for realistically large numbers of word groups). Since several ants are used in each epoch and the solution

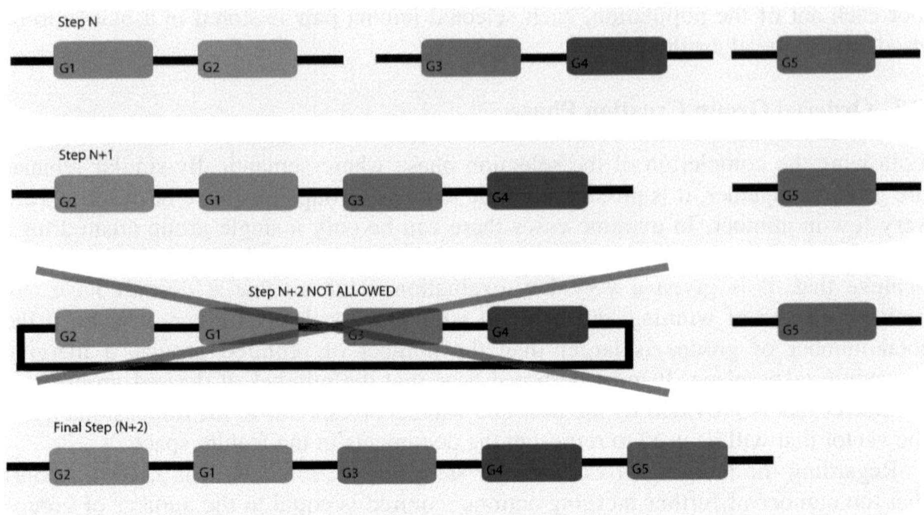

Fig. 4. Gradual ordering of the groups on the one-dimensional structure. Note that in step N+2, the most similar groups are G4 and G2 but these are not allowed to be connected as they would result in a circular rather than open-ended one-dimensional structure.

is revised in each epoch only according to the iteration-best ant, it is sufficient to find one good solution within each iteration. Furthermore, the gradual optimization process of the ACO approach through several iterations allows the system to recover even from iterations where no good solutions whatsoever are generated.

3.4 Solution Evaluation Phase

By utilizing this list structure, it is possible to calculate a similarity between lemmas according to the groups they have been assigned to. In the proposed system, lemmas that belong to the same group are considered to have a similarity inversely proportional to the size of the group they belong to. For lemmas that belong to different groups, the similarity is inversely proportional to the sum of the group sizes that need to be traversed when moving from the group of the one lemma to the group of the other (including the source and target groups). This group size factor was introduced in order to increase the similarity measure between lemmas that belong to the smaller groups, since smaller groups are more likely to represent more concise and similar concepts than larger ones. Additionally, similarities between groups should be stronger when smaller groups are situated in between, since larger groups tend to attract more lemmas and thus represent more generic information. This similarity measure between lemmas is used to evaluate each individual ant (i.e. the corresponding solution) of the ACO procedure and to update the pheromone trails accordingly.

Initially, in order to speed up execution, all similarities among groups are cached. Then, for each period within the text corpus, the total similarity between lemmas is calculated based on the group list created. This similarity is defined by adding the similarity of each lemma in the period with every other lemma. For the estimation of the total similarity in a period, weights are used that are inversely proportional to the distance in terms of words in the period (smaller weight values correspond to more distant lemmas). These weights were introduced in order to emphasize similarities on lemmas that are direct neighbors. The total value of the evaluation function is obtained by adding the in-period similarities of all the periods in the dataset. The complete form of the evaluation function is presented in equation (3):

$$E_m = \sum_{s=1}^{Sentences} \sum_{i=1}^{ns} \sum_{j=1 \& i \neq j}^{ns} (sim(w_i, w_j) \cdot f^{|i-j|-1}) \tag{3}$$

In equation (3), m is the identifier of the selected group and E_m is the value of the evaluation function. The ns parameter corresponds to the number of words in the sentence and w_i indicates the ith word. The similarity between the ith and the jth words, denoted by $sim(w_i, w_j)$, is estimated on the basis of the grouping. f is a positive value ($0 < f < 1$) that lowers the similarity between words in the sentence as their distance in terms of words ($|i-j|$) increases.

Based on the evaluation function, only the best iteration individual (ant) is selected for updating the pheromone on the trail paths. This option has been preferred over a proportional update where all individuals would take part, since it is much faster to execute. All pheromone trail strengths are initially reduced by an evaporating factor, while only the trails (which in the presented case are equivalent to word pairs) that correspond to the solution of the best individual are reinforced by a constant value. The total value of the pheromone in the system remains constant. While the system advances through iterations, the best solution is saved but no elitism is adopted and thus no individual is copied to the next generation. Instead, the ACO update scheme follows an iteration-best approach where the pheromone is updated only to record the trails of the best ant of the current iteration. It must be noted here that since the construction of word groups involves a random selection step based on pheromone trails and heuristic information, it is quite possible that even when processing the same individual the final word grouping can be totally different.

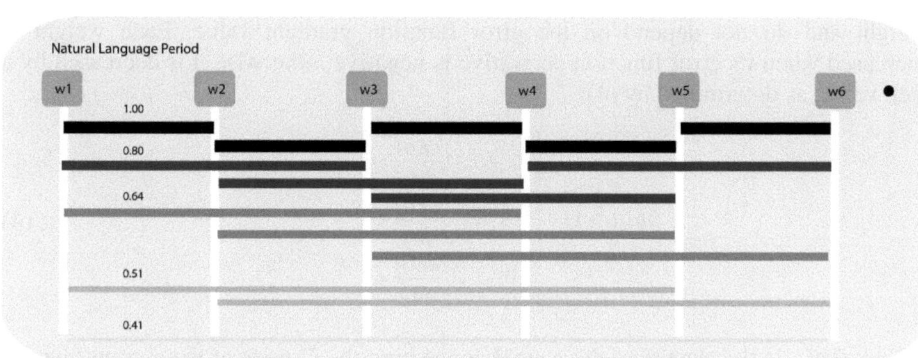

Fig. 5. Sentence evaluation with weights and distances

4 Classification Module

The grouping of words/lemmas generated by the word map module is subsequently used to describe the documents of the dataset. For every document, the number of lemmas that belong to each group is counted. This set of measurements forms the representative vector of a document, which has a dimensionality equal to the number of groups selected. Each vector is normalized so that the sum of its components is equal to one and each element reflects the frequency of lemmas from a given group.

To classify the documents an MLP-based module is used. For each document the corresponding vector of numeric values is provided as the input to the MLP. The number of input nodes of the MLP is equal to the vector dimensionality (equal to the numbers of groups created from the ACO - see Fig. 4). The assignment of the groups on the one-dimensional structure is not taken into account by the MLP, which only focuses on the words assigned to a given group. The MLP output nodes are set equal to the number of required classes of documents. The number of hidden nodes is experimentally determined (see section 6). In order to accurately estimate the classification accuracy achieved by the MLP, the leave-one-out evaluation scheme has been chosen.

The MLP consists of layers of neurons, where each neuron is stimulated by the weighted sum of the outputs of previous-layer neurons (Rumelhart et al. 1986). Every neuron processes its activation signal through a non-linear bipolar activation function (such as the hyperbolic tangent function) and transmits the output signal to the next-layer neurons. It has been reported (Haykin 1999) that MLPs are capable of creating any input-to-output (i.e. from features to decision) mapping, by using a single hidden layer that contains an adequate number of neurons.

The most widely used method for the supervised training of an MLP is via the family of backpropagation algorithms (Rumelhart et al. 1986). In backpropagation, the deviation of the network's output from a desired target is measured via the mean squared error. The backpropagation algorithm determines the network error at the output layer and then propagates the error signal from the output towards the input, one layer at a time.

The RPROP algorithm (Igel & Huesken 2003 and Riedmiller & Braun 1993) adapts independently weights in predetermined steps, which are different for each weight and do not depend on the error function gradient value. Each weight is increased when its error function derivative is negative; otherwise it is decreased by a step value, as determined by (4):

$$\Delta w_{ij}(t+1) = \begin{cases} -g_{ij}(t), & \dfrac{dE(t)}{dw_{ij}} > 0 \\ +g_{ij}(t), & \dfrac{dE(t)}{dw_{ij}} < 0 \\ 0, & \dfrac{dE(t)}{dw_{ij}} = 0 \end{cases} \tag{4}$$

where Δw_{ij} is the change of the weight connecting the ith node of the current layer to the jth neuron of the next layer and g_{ij} is the update step of weight w_{ij}. When the error function derivative (gradient) with respect to the weight retains the same sign for a consecutive number of training steps, the step value (g_{ij}) is progressively increased to make the network converge faster. On the other hand, when consecutive sign changes are observed, the weight step value is decreased to make the procedure convergence more accurate near local minima, as expressed by equation (5):

$$g_{ij}(t+1)=\begin{cases} n^{+}\cdot g_{ij}(t), & \frac{dE(t)}{dw_{ij}}\cdot\frac{dE(t\text{-}1)}{dw_{ij}}>0 \\ n^{-}\cdot g_{ij}(t), & \frac{dE(t)}{dw_{ij}}\cdot\frac{dE(t\text{-}1)}{dw_{ij}}<0 \\ 0, & \frac{dE(t)}{dw_{ij}}\cdot\frac{dE(t\text{-}1)}{dw_{ij}}=0 \end{cases} \tag{5}$$

where n^{+} is a step-increase value (with $n^{+} > 1$) and n^{-} is a step-decrease value (with $0 < n^{-} < 1$).

The chosen MLP architecture comprises three layers. The input layer contains k neurons, where k is equal to the number of word groups generated by the SOM. The output layer contains a number of neurons equal to the document categories. In order to improve the training, the Nguyen – Windrow initialization technique is adopted (Nguyen & Widrow 1990), ensuring that initial weights are uniformly distributed in the input space.

5 Alternative Implementation Using a SOM-Based System

In order to verify the proposed ACO system its performance was compared to the SOM-based system that was proposed by (Tsimboukakis & Tambouratzis 2007). In that system, lemmas are first organized into groups on a Self Organizing Map (Kohonen, 1997) which serves a function similar to the ACO-based system. The dimensionality of the feature vector used as input to the SOM is significantly lower than the number of lemmas available in the dataset. For each lemma, the ideal case would be to record the number of co-appearances with all lemmas in the dataset. Such a solution would highly increase the dimensionality of the lemma feature vectors. To avoid the increase in dimensionality, only a subset of available lemmas was used (Tsimboukakis & Tambouratzis 2007) as the feature set, so that every lemma could be described by its co-occurrences with the feature set lemmas.

The selection of feature lemmas was made via an ad-hoc rule. The frequency of appearance for each lemma in the whole corpus is assumed to follow Zipf's distribution, where there are only a few lemmas that are extremely frequent and many lemmas that are very rare. As an analogy to Zipf's law, Pareto's principle (also known as the 80-20 rule) states that 20% of causes are responsible for the 80% of the results. Pareto's principle, also termed the ABC analysis, has been mostly applied to quality control and management tasks. According to the ABC analysis, a portion of the causes is characterized as belonging to category A which indicates very important events, with B and C corresponding to less important and to unimportant events respectively. In this case category A contains highly-frequent lemmas (which correspond to functional words, such as articles, auxiliary verbs and conjunctions), B contains frequently-used lemmas and C contains rare lemmas. It seems appropriate to select the feature lemmas from category B, since lemmas within this frequency range do not correspond to functional or very common words (which do not reflect a specialized content) and yet are frequent enough to describe the remaining lemmas. The limits of the ABC analysis have been chosen as follows:

- category A contains the most frequent lemmas that collectively amount to 70% of all appearances.
- category B contains lemmas that contribute the next 15% (70-85%) of appearances.
- category C contains lemmas that correspond to the remaining 15% of appearances.

Furthermore, lemmas that appear less than three times are omitted from the remainder of the analysis, as their frequency of occurrence is too low to provide measurable frequencies. The context-based frequencies of these lemmas (categories A and C) with respect to the feature lemmas of Category B are used to cluster the documents.

More specifically, every lemma w_i is represented by a vector of elements, where each element corresponds to one B-lemma and indicates the number of times the given lemma w_i occurs in the same period as that B-lemma within the document dataset. It must be noted that only sentence-level appearances are counted and if a lemma or a feature appears more than once within a sentence it is only counted once. The use of periods between major punctuation marks as a basis for counting the appearances of lemmas is based on the principle that a full-stop between sentences is the least ambiguous point at which the description of an idea or event is terminated, while in subsequent sentences it is likely that completely unrelated concepts are presented. The components for each vector of word co-appearances are defined by the following equation:

$$s_f(w_i) = \frac{N(w_i,f)}{N(w_i)} \tag{6}$$

where $s_f(w_i)$ is the measurement for feature f of lemma w_i, $N(w_i,f)$ is the number of sentences where both lemma w_i and feature f appear together and $N(w_i)$ is the number of sentences where lemma w_i appears. Each vector is normalized so that its components sum to one, similarly to the approach followed for the ACO-based process as described in section 3, except from the fact that no weighting factor is adopted for the distance between words in the natural language period.

Since each lemma can be described by a numeric vector, it is straightforward to use a common clustering algorithm such as the Self-Organizing Map to group lemmas that appear in similar contexts, and thus possess similar vectors. It is reasonable to assume that words which appear in similar contexts bear related meanings. SOM models have been applied to word grouping tasks by other researchers (Kohonen & Somervuo 1998 and Georgakis et al. 2004).

In brief, SOM networks usually comprise a single-layer structure of neurons arranged into a regular structure, while each neuron is connected to every input. Each of the neurons contains a weight vector with dimensionality equal to the input dimensionality.

The SOM model operates in a competitive manner, the output of the training process being the identity of the neuron that best matches the pattern presented, in terms of the predefined distance. This winning neuron is then adapted towards the training pattern. The SOM neurons are connected with their direct neighbours depending on the lattice geometry selected. The most commonly-used lattice is the 2-dimensional hexagonal structure as it approximates more closely a uniform distribution of neurons, this structure being used in the experiments described here.

The SOM training process is unsupervised and thus desired outputs for input patterns are not provided. The neuron weights are initialised linearly along the two eigenvectors of the training data autocorrelation matrix that correspond to the two largest eigenvalues. Following initialisation, every pattern is presented to the network and the winner neuron is adapted so as to increase its similarity to the pattern. The neighbouring lattice neurons are also updated to a decreasing degree as their distances from the winner increase. The weight update function used to update neurons is a Gaussian kernel function of the distance on the lattice:

$$\Delta w_i = \eta(n) \cdot h_{i,c(x)}(x - w_i) \tag{7}$$

where Δw_i is the weight change for the ith neuron, η is the learning rate, h is the neighbourhood function and $c(x)$ is the winner neuron for training pattern x.

In order to stabilise the learning procedure, the batch training algorithm allows weight updates to only take place after the entire training set is presented. Then each neuron weight update equals the average of the data vectors within its neighbourhood:

$$w_i = \frac{\sum_j h_{i,c(x_j)} \cdot x_j}{\sum_j h_{i,c(x_j)}} \tag{8}$$

The neighbourhood radius is decreased during training, so that the SOM network finally converges to a stable state. For the experiments reported here, an environment in C++ was developed, this being a port of the original SOM toolbox (Vesanto 2000) developed for MATLAB.

The network size used was automatically determined from the data using the estimation procedure provided by the SOM toolbox, the map comprising 22x19 neurons. Lemmas that were assigned to category B of the ABC analysis were omitted from the map, since they are present in the feature vectors. Hence, the B-lemmas give very high values for the corresponding feature (since $s_f(w_i = f) = 1$ from equation (6)) and they could influence heavily the training algorithm, substantially more than the other lemmas.

When using a SOM map to cluster data, the number of nodes is frequently chosen to be higher than the expected number of clusters, due to the fact that the clustering method is unsupervised and thus it is likely that multiple neighboring nodes correspond to the same actual class. Since the number of groups that are input to the next stage of the system (the classifier stage) influences highly the classifier complexity, further compression of the map would be desirable for the document classification task. This is achieved by adding a batch k-means step. The batch k-means (Duda et al. 2001) algorithm is initialized evenly with respect to the 2-D word map, so that the distance on the grid between two neighboring centers at each dimension is almost constant. Initial k-means centers are placed on the vertexes of a rectangular lattice, where each of the two edges of the rectangular lattice has a length proportional to the SOM lattice size.

Each initial center is given the value of the nearest neuron's vector, and thus has a dimensionality equal to those of SOM neurons. The SOM grid is actually taken into account only at the initialization step, while the distance between centers is calculated from their vectors. Each iteration of the algorithm consists of two distinct phases, where initially (i) every codebook vector is assigned to the nearest k-means center and then (ii) each center is updated to the mean of the vectors assigned to it:

$$\vec{c}_i(t+1) = \frac{1}{N}\sum_{j=1}^{N}\vec{d}_j \qquad (9)$$

where $\vec{c}_i(t+1)$ is the center i at iteration $t+1$, N is the number of patterns that were assigned to $\vec{c}_i(t)$ and $\vec{d}_j$ is the jth pattern assigned to $\vec{c}_i(t)$.

Following the dimensionality reduction process, *k* groups of lemmas are generated. The value of *k* corresponds to the level of complexity and detail that the user requires from the clustering system.

6 Experimental Results

In order to assess the effectiveness of the proposed ACO-based solution, an implementation of the ACO was made in the C++ programming language. Then the system was applied to a real-world classification problem. The dataset used for illustrating the behavior of the proposed system comprises texts from the Minutes of the Greek Parliament. These Minutes contain a large amount of texts spanning almost two centuries, edited via a well-established procedure by specialized personnel and readily available in electronic format. From the Minutes, a set of documents have been collected which comprise a total of 5 speakers for a given Parliamentary period (a 4-year period between two consecutive general elections). The full dataset contains 1,004 documents, which have a document length varying from just over 100 up to 8,000 words. These documents have been independently assigned by specialized linguists to categories on the basis of their content, as shown in Table 1. Out of these texts, a total of 560 documents have been assigned to one of the four most popular categories, namely internal affairs, external affairs, education, and economy. The problem that the proposed system is employed to solve is the assignment of previously unseen documents to one of the available categories with the highest possible accuracy.

The ACO-based system comprises a number of parameters whose modification is likely to affect its performance. Some of these parameters were imposed by the need to render the simulation of the ACO system for a sufficient number of iterations computationally feasible. To that end, very frequent lemmas that appeared more than a given number of times were removed since these words would appear more or less in every document and thus they would not provide any useful information in terms of

Table 1. Distribution of documents per category in the data set

Category	Number of documents
Internal Affairs	82
External Affairs	207
Education	70
Economy	201
Other Categories	444
Total	1,004

document topic. The values chosen for the frequent word threshold were 1000 and 5000 (for the given collection of 1,004 documents). Along with frequent lemmas, functional words were also ignored. Furthermore, rare lemmas that had only few appearances within the dataset were removed since they add further complexity to the system without providing important information. The two threshold values selected in the experiments for the removal of rare words were 5 and 10. Sentences that contained more than 50 words were not taken into account since these sentences may disproportionately affect the results of the evaluation. The proximity parameter f of equation (3) denotes the degree to which the similarity between two words in the specific period is counted to the final result, based on their distance in this period, and it was set to 0.8. The evaporating factor that denotes the level of pheromone reduction after each epoch was selected to be equal to 0.97 and the population size was set to 30 ants. It would be desirable to have greater population sizes but this would substantially reduce the number of epochs that could be completed within a total simulation period for the proposed ACO implementation. The number of groups required by the algorithm was equal to 50, which is adequate for the given dataset as previously experiments with different systems have shown (Tsimboukakis & Tambouratzis 2007 and Tsimboukakis & Tambouratzis 2008).

An additional parameter of the proposed architecture is the size of the hidden layer of the MLP. Since a deterministic procedure is not available for defining the number of hidden-layer neurons, this has been determined by experiments with network sizes ranging from 5 to 10 hidden neurons.

For the 560 documents of the dataset that belong to one of the four topic categories, nine subsets were created. Seven of these were used to form the training set, one subset as validation set for the appropriate termination of the MLP training and the final subset as the test set for evaluating the system performance. All possible combinations of sets were run and the average performance on the test set is reported. The results obtained for the various system setups corresponding to different word removal thresholds along with different MLP sizes are reported in Table 2. All results quoted here correspond to a simulation of the ACO-based system, for a total of 25 iterations.

The classification accuracy observed for the four-topic discrimination task varies between 60.7% and 69.4% for all experiments performed. The maximum accuracy observed was 69.4% for an MLP with 7 hidden neurons when words that had more than 1000 and less than 5 appearances were removed (typically, starting with 7,600 distinct lemmas within the data set, introducing the upper and lower limits in the frequencies-of-appearance, a total of 6,142 lemmas are finally retained). It can be

Table 2. Classification accuracy for the classification task involving four topic categories, reported for various MLP sizes

ACO Parameters	MLP 5	MLP 6	MLP 7	MLP 8	MLP 9	MLP 10	Lemmas Retained
Freq Thresh. 1000 Rare Thresh. 10	68.2%	67.8%	68.2%	69.2%	69.2%	69.1%	4,461
Freq Thresh. 5000 Rare Thresh. 10	61.3%	61.0%	60.7%	60.7%	61.2%	62.6%	4,551
Freq Thresh. 1000 Rare Thresh. 5	66.5%	68.7%	69.4%	68.7%	69.2%	68.9%	6,142
Freq Thresh. 5000 Rare Thresh. 5	66.3%	67.4%	65.8%	67.3%	68.2%	67.9%	6,232

seen from Table 2 that varying the MLP size affects the performance by less than 2%. It appears that systems which use an upper threshold of 1,000 for eliminating extremely frequent words always perform better that those that use a threshold equal to 5,000. This fact indicates that when more frequent words are removed, the system tends to perform better and thus the removal of frequent words proves beneficial. On the other hand, the value of the threshold for rare word removal does not appear to have a consistently positive or negative effect to the classification accuracy. According to Table 2, the selection of thresholds equal to 10 and 5000 negatively affects the ACO-based system, this combination corresponding to the second line on table 2. This configuration exhibits the worse performance than any other ACO-based system setup, resulting in an accuracy reduction of at least 5%.

Unfortunately neither system setup does reach the classification accuracy observed by the alternative SOM-based system which was reported to be 74% (Tsimboukakis & Tambouratzis 2007). The best classification accuracy observed using ACO-based methods remains 4.6% lower than the SOM-based one.

Another general comment involves the number of hidden nodes in the network. According to the experimental results of Table 2, smaller networks (for instance networks with 5 hidden nodes) result in a lower accuracy, while networks with more than 7 hidden nodes illustrate a higher accuracy, in general.

Due to the fact that topic classification can be ambiguous, depending on each individual's background and preferences, a second experiment was also carried out. From the topic categories, the most general category ("Internal Affairs") was removed since there are numerous cases where documents from this category also belong to one of the other, more specific categories. Then, the same experiments were run for the three-category classification task, the results being presented in Table 3.

According to Table 3, following the reduction in the number of categories, the classification accuracy of the ACO-based system is increased by more than 10% in all cases. This is similar to the observation when using the SOM-based document classification system (Tsimboukakis & Tambouratzis 2007) for the three-category classification task. The best classification accuracy observed is 81.3% for the ACO-based system, when using upper and lower frequency thresholds of 1,000 and 10, respectively. The four configurations tested (which correspond to the four rows of Tables 2 and 3) tend to keep the same ranking in terms of performance for both the three-category and the four-category classification tasks. The best system

Table 3. Classification Accuracy for the three-category classification task for various MLP sizes

ACO Parameters	MLP 5	MLP 6	MLP 7	MLP 8	MLP 9	MLP 10	Lemmas Retained
Freq Thresh. 1000 Rare Thresh. 10	79.9%	81.3%	80.5%	81.0%	80.5%	79.6%	4,461
Freq Thresh. 5000 Rare Thresh. 10	73.4%	74.1%	73.8%	74.4%	74.2%	74.1%	4,551
Freq Thresh. 1000 Rare Thresh. 5	77.6%	79.7%	79.0%	77.7%	79.4%	79.5%	6,142
Freq Thresh. 5000 Rare Thresh. 5	78.7%	78.3%	79.4%	79.2%	78.8%	78.5%	6,232

setup remains that with values 1,000 and 10. In the case of the three-category classification task, the MLP size doesn't seem to affect substantially the classification accuracy, indicating that even an MLP network with 5 hidden nodes suffices for the 3-category classification task. The classification accuracy obtained by the alternative SOM system reaches 84% which is 2.7% larger than the proposed ACO implementation.

In figure 6, the variance of the pheromone trails value is displayed throughout the ACO system adaptation iterations. It can be noted that the variance increases as expected, which denotes that the ACO system starts to develop preferences over specific word connections (paths) rather than others. From figure 6 it can be seen that more iterations of the ACO procedure could prove valuable as the variance still increases with almost linear rate at the point where the simulation is terminated.

This observation is actually of substantial interest. The proposed system was simulated on a MS-Windows based system, occupying a single processor of a

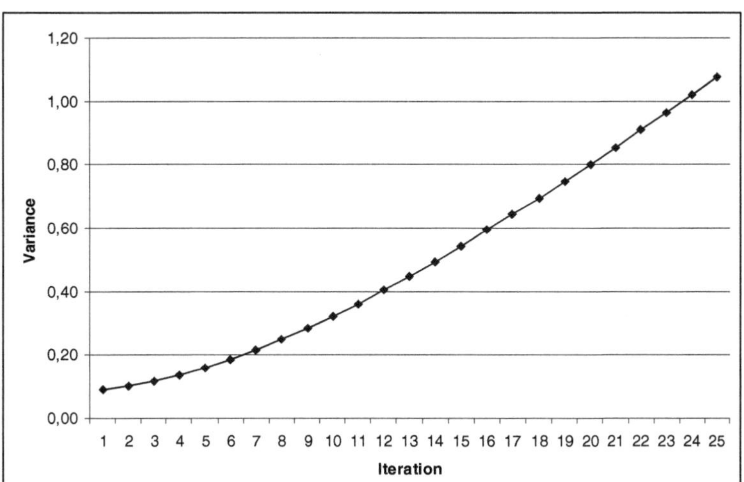

Fig. 6. Evolution of the variance of the pheromone distribution throughout a typical ACO simulation

quad-core system using an Intel Q6600 processor clocked at 2.66 Mhz. Throughout the implementation of the ACO-system, care was taken to minimize the processing requirement, in many cases choosing approximate solutions to reduce the processing requirements. Still, it has been found that approximately 14 days are required to simulate a total of 25 epochs of the ACO algorithm, for a population of 30 ants. In this respect, it is expected that improvements to the ACO algorithm (for instance, Visine et al., 2005 and Chen & Li, 2007) can be employed to improve the system effectiveness and speed of convergence.

7 Conclusion

Document organization systems are extremely important for informational retrieval and storage applications in large organizations and enterprises. In the present study a hybrid ACO algorithm variation has been presented that has been applied in a document classification system for Greek documents. The majority of document organization systems tend to adopt lemma-based representation of documents in the feature space. This is the case in the well-known WEBSOM (Kohonen et al. 2000 and Lagus et al. 2004) for example, which operates by representing each document with a number of TF-IDF features (Drucker et al. 1999). On the other hand the system presented here employs an intermediate word map that aims at creating groups of lemmas with related meanings. The thematic relation is detected by the frequent co-appearance of words in natural language periods. The creation of this map is implemented by a hybrid ACO algorithm and its effectiveness is compared to that of a SOM neural network.

According to the experimental results, the proposed system requires more CPU time for adaptation than the SOM system for training and doesn't surpass the classification accuracy of the SOM. On the other hand, the ACO-based system proposes a novel solution for encoding symbolic information such as the word context in natural language periods. Additionally the group creation is handled in a completely different manner to the SOM-based system. An obvious advantage is that the ACO system doesn't need a confined subset of lemmas (such as the B category words of the SOM-based system) to represent other lemmas. Thereby, important word occurrences such as the B words are not ignored from the document representation vectors, as was the case in the SOM-based system. Moreover the ACO system is more suited to dynamically changing environments where large amounts of new documents may be introduced after the partial training of the system, and thus new lemmas may appear that can prove in time to be frequent - for instance in the case when a new topic category is introduced to an existing system. The ACO system presented here would require more extensive experimentation and tuning of parameters, while implementation improvements especially in the selection phase of the ACO algorithm would be necessary. Finally the proposed ACO-based system could certainly benefit from the parallelization of the algorithm, which would allow it to improve the execution times by utilizing the processing power of multiple CPUs.

References

1. Chen, X., Li, Y.: A Modified PSO Structure Resulting in High Exploration Ability With Convergence Guaranteed. IEEE Transactions on Systems, Man & Cybernetics, Part B: Cybernetics 37(5), 1271–1289 (2007)
2. Dorigo, M., Gambardella, M.: Ant Colony System: A Cooperative Learning Approach to the Travelling Salesman Problem. IEEE Transactions on Evolutionary Computation 1(1), 53–66 (1997)
3. Drucker, H., Wu, D., Vapnik, V.: Support Vector Machines for Spam categorization. IEEE Trans. on Neural Networks 10(5), 1048–1054 (1999)
4. Dussutour, A., Fourcassie, V., Helbing, D., Deneubourg, J.-L.: Optimal Traffic Organisation in Ants under Crowded Conditions. Nature 478(6978), 70–73 (2004)
5. Duda, R.O., Hart, P.E., Stork, D.G.: Pattern Classification, 2nd edn. Wiley, New York (2001)
6. Everitt, B.S., Landau, S., Leese, M.: Cluster Analysis. Hodder Arnold Publication (2001)
7. Georgakis, A., Kotropoulos, C., Xafopoulos, A., Pitas, I.: Marginal median SOM for document organization and retrieval. Neural Networks 17(3), 365–377 (2004)
8. Freeman, R.T., Yin, H.: Web content management by self-organization. IEEE Transactions on Neural Networks 16(5), 1256–1268 (2005)
9. Haykin, S.: NEURAL NETWORKS: A Comprehensive foundation, 2nd edn. Prentice-Hall, Englewood Cliffs (1999)
10. Igel, C., Huesken, M.: Empirical evaluation of the improved Rprop learning algorithm. Neurocomputing 50, 105–123 (2003)
11. Kaski, S.: Dimensionality Reduction by Random mapping: Fast Similarity Computation for Clustering. In: Proceedings of IJCNN 1998 Conference, International Joint Conference on Neural Networks, vol. 1, pp. 413–418 (1998)
12. Kennedy, J., Eberhart, R.C.: Swarm Intelligence. Morgan Kaufmann, San Francisco (2001)
13. Kohonen, T.: Self-Organizing Map, 2nd edn. Springer, Berlin (1997)
14. Kohonen, T., Somervuo, P.: Self-organizing maps of symbol strings. Neurocomputing 21(1-3), 19–30 (1998)
15. Kohonen, T., Kaski, S., Lagus, K., Salojarvi, H.J., Patero, V., Saarela, A.: Self Organisation of a Massive Document Collection. IEEE Transactions on Neural Networks 11(3), 574–585 (2000)
16. Lagus, K., Kaski, S., Kohonen, T.: Mining Massive Document Collections by the WEBSOM Method. Information Sciences 163(1-3), 135–156 (2004)
17. Lessing, L., Dumitrescu, I., Stützle, T.: A comparison between ACO algorithms for the set covering problem. In: Dorigo, M., Birattari, M., Blum, C., Gambardella, L.M., Mondada, F., Stützle, T. (eds.) ANTS 2004. LNCS, vol. 3172, pp. 1–12. Springer, Heidelberg (2004)
18. MacKay, D.: Information Theory, Inference, and Learning Algorithms. Cambridge University Press, Cambridge (2003)
19. Manning, C.D., Schütze, H.: Foundations of Statistical Natural Language Processing. MIT Press, Cambridge (1999)
20. Martens, D., De Backer, M., Haesen, R., Baesens, B., Mues, C., Vanthienen, J.: Ant-Based Approach to the Knowledge Fusion Problem. In: Dorigo, M., Gambardella, L.M., Birattari, M., Martinoli, A., Poli, R., Stützle, T. (eds.) ANTS 2006. LNCS, vol. 4150, pp. 84–95. Springer, Heidelberg (2006)
21. Nguyen, D., Widrow, B.: Improving the learning speed of 2-layer neural networks by choosing initial values of adaptive weights. In: Proceedings of IJCNN 1990 Conference: International Joint Conference on Neural Networks, vol. 3, pp. 21–26 (1990)

22. Papageorgiou, H., Prokopidis, P., Giouli, V., Piperidis, S.: A Unified PoS Tagging Architecture and its Application to Greek. In: Second International Conference on Language Resources and Evaluation Proceedings, Athens, Greece, vol. 3, pp. 1455–1462 (2000)
23. Riedmiller, M., Braun, H.: A direct adaptive method for faster backpropagation learning: the RPROP algorithm. In: Proceedings of the IEEE International Conference on Neural Networks, San Francisco, CA, pp. 586–591 (1993)
24. Rumelhart, D.E., Hinton, G.E., Williams, R.J.: Learning representations by back-propagating errors. Nature 323, 533–536 (1986)
25. Salton, G., Buckley, C.: Term-weighting approaches in automatic text retrieval. Information Processing & Management 24(5), 513–523 (1988)
26. Stützle, T., Hoos, H.: MAX-MIN Ant System. Future Generation Computer Systems 16, 889–914 (2000)
27. Tsimboukakis, N., Tambouratzis, G.: Self-Organizing Word Map for Context-Based Document Classification. In: Proceedings of the WSOM 2007 International Workshop on Self-Organizing Maps, Bielefeld, Germany (2007)
28. Tsimboukakis, N., Tambouratzis, G.: Document classification system based on HMM word map. In: CSTST 2008, 5th International Conference on Soft Computing as Transdisciplinary Science and Technology, Paris, France, pp. 7–12 (2008)
29. Tweedie, F., Singh, S., Holmes, D.: An Introduction to Neural Networks in Stylometry. Research in Humanities Computing 5, 249–263 (1996)
30. Vesanto, J.: Neural Network Tool for Data Mining: SOM Toolbox. In: Proceedings of Symposium on Tool Environments and Development Methods for Intelligent Systems (TOOLMET 2000), Finland, pp. 184–196 (2000)
31. Visine, A.L., de Castro, L.N., Hruschka, E.R., Gudwin, R.R.: Towards Improving Clustering Ants: An Adaptive Ant Clustering Algorithm. Informatica 25, 143–154 (2005)

12

Identifying Disease-Related Biomarkers by Studying Social Networks of Genes

Mohammed Alshalalfa, Ala Qabaja, Reda Alhajj, and Jon Rokne

Department of Computer Science, University of Calgary, Calgary, Alberta, Canada
Department of Computer Science, Global University, Beirut, Lebanon
msalshal@ucalgary.ca, alhajj@ucalgary.ca, rokne@ucalgary.ca

Abstract. Identifying cancer biomarkers is an essential research problem that has attracted the attention of several research groups over the past decades. The main target is to find the most informative genes for predicting cancer cases, such genes are called cancer biomarkers. In this chapter, we contribute to the literature a new methodology that analysis the communities of genes to identify the most representative ones to be considered as biomarkers. The proposed methodology employs iterative t-test and singular value decomposition in order to produce the communities of genes which are analyzed further to identify the most prominent gene within each community; the latter genes are analyzed further as cancer biomarkers. The proposed methods have been applied on three microarray datasets. The reported results demonstrate the applicability and effectiveness of the proposed methodology.

1 The Motivation and Contributions

As part of finding the proper cure for diseases, understanding the molecular mechanism of diseases is a major concern for biologists and medical scientists. Cancer is one of the major diseases threatening human life. In the United States, for instance, cancer is number one cause of death males and number two in women after cardiovascular diseases [1]. Cancer is considered as the most challenging disease ever as there is no particular confirmed reason for it. In addition, cancerous cells exhibit different behavior and have different mechanisms for developing and deviating from the normal behavior. Early cancer diagnosis is a critical step to cure cancer cases as it has already been shown easier to treat people diagnosed at early stages than those who are diagnosed late; the former cases are mostly controllable.

Different techniques, like Northern blotting and Real-Time Polymerase Chain Reaction (RT-PCR), are used [2] to measure the expression level of genes in cancer cells. Using such techniques is time consuming, error prone and expensive as they can measure the expression level of at most 50 genes at a time. Thanks to

C.P. Lim et al. (Eds.): Innovations in Swarm Intelligence, SCI 248, pp. 237–253.
springerlink.com

the microarray technology, where up to 10,000 genes can be studied simultaneously under different conditions. The microarray technology gets its popularity among molecular biologists because it allows studying the transcriptome of organisms simultaneously under different conditions; a transcriptome is a set of the gene transcripts present in a given cell.

Data mining and statistical techniques made it easier to interpret, understand, and extract the knowledge hidden within the microarray data. In other words, to analyze the expression level of all the genes in human cancer, microarray is used to study the expression level of all the genes in "normal" and cancerous humans. The result of gene expression study is a matrix of m genes and n samples, where samples represent either "normal" or cancerous humans; this matrix is called the microarray data. The goal to be achieved is the ability to distinguish between normal and cancer samples based on subset of features (genes) selected from microarray data. To help biologists and medical scientist developing effective analysis, different statistical and computing techniques are employed in the process; the main target is to reduce the space for better control and analysis. In this sense, clustering algorithms, *e.g.*, k-means, Fuzzy C-means [3] and Self-Organizing Maps (SOM) [4] have been used to cluster the samples into two or more classes depends on the number of available cancer samples [5].

While clustering is known as unsupervised learning technique capable of discovering the classes, using supervised learning techniques is preferred when the classes are known. Thus, more efficient techniques, like Support Vector Machine (SVM) [6] and Neural Networks (NN) [7], have been effectively employed to classify the samples. The process applied uses the knowledge in part of the data to build a classifier model, which is then tested on the other part of the data.

Applying clustering or classification techniques directly to microarray data showed to be very poor mainly because microarray data is noisy and has redundant and insignificant information. Therefore, it is necessary to incorporate a preprocessing step to eliminate redundant and insignificant information from microarray data. Such process is intended to eliminate genes that do not contribute to cancer; that is, to eliminate genes that do not change their expression in cancer cells, as they are irrelevant to cancer. So, the problem tackled in this chapter is to develop techniques effective in extracting the biomarker genes that contribute mostly to cancer. The biomarker genes should be able to correctly distinguish between normal samples and cancer samples, and the number of extracted biomarker genes should be small. Another objective to be considered is to reduce the functional redundancy of the extracted genes. We need to get representative genes from each pathway because cancer is caused by different molecular pathways in the cell like P53, AKT and EGF pathways. Actually, genes interactions are realized into social network to be studied and analyzed for better discoveries. Discovering the social communities of genes lead to eliminating redundancy and hence produces a more compact and mostly optimized set of representative genes that could be studied further as cancer biomarkers. The latter genes are generally the most significant within their social communities. To produce a workable and effective solution, we want to benefit from machine learning and statistical techniques to extract the significant biomarker genes, which are then to be used for building a classifier model. The accuracy of the classifier will be used to evaluate the goodness of the extracted genes. Finally, we want to ensure high statistical significance of the extracted genes. T-test and Analysis of Variance (ANOVA) are the the most widely used methods to ensure high statistical significant of microarray data, but they require a cut-off value. Additionally, they are individual gene based methods, *i.e.*, they do not consider the relationship among genes. All of these gaps present in the literature motivated the approach proposed in this book chapter. We need a robust framework that derives and studies social networks of genes to better identify cancer biomarkers.

1.1 The Proposed Approach

In a previous study [8,9,10,11], we realized the advantage of clustering in analyzing social communities of genes; we grouped genes with similar gene expression profile into clusters, where every cluster is represented by its centroid. The intention here is to use the gene closest to the centroid as reduced feature to represent the cluster. Thus, after the clustering is over, the genes closest to centroids (one gene per cluster) represent the whole data. Since the clusters may have large number of genes, and the effectiveness of a single representative per cluster is directly affected by the level of homogeneity of each cluster, we produced more representatives per cluster by applying a second clustering on each cluster from the first clustering. In other words, clustering the clusters into subclusters generally produced more representative centroids. The obtained centroids have high potential to distinguish between classes after noise elimination. The clustering of subclusters may recursively propagate as there is need for more representative genes; the upper limit is a structure similar to hierarchical clustering. However, we do not expect to go beyond the third level if ever needed; our results reported in the literature [8,9,10,11] demonstrate that two levels of clustering would lead to the required number of representative genes.

Though the clustering based approach reported good results, there are other perspectives that might help developing other approaches. So, we tried to investigate other alternative approaches that might also help us to validate the produced results by considering some aspects of the gene expression data. The rational for two other approaches is described in the next two paragraphs, respectively. These two other approaches involve more biological information into the process. At the end, having three approaches will put us on a more solid ground as we analyze the reported significant genes because the participation of genes to social communities may overlap; in other words, the same gene may be identified by different approaches as members of different social communities. Hence, it is necessary to produce a compromise and carefully analyze the actual social community to which each gene should belong. The later step is left as future work, and the ultimate target is to integrate the three approaches into a robust framework.

Ranking genes based on their entropy is considered a promising method proposed by Varshavsky *et al.* [12], but the algorithm they proposed was designed for time series data gene selection, not for sample classification data. The proposed method was adapted in this chapter to select significant genes in social communities produced from classification data. Both statistical tests and singular Value Decomposition (SVD) based approaches are deficient. T-test only considers whether the gene expression level in normal samples is significantly different from cancer samples. They do not consider the other genes while ranking, and they do not extract genes which have very close gene expression level among samples. In other words, the disadvantage of the statistical test is to set the threshold value required to select the top genes. Also, redundant genes can not be eliminated using these statistical tests; these methods do not consider the whole data. They just evaluate, within the social community of genes, the significance of each gene individually, and select the top ones, depending on the threshold. On the other hand, SVD [12] considers the whole data and assigns a weight to each gene. SVD ranks genes based on their effect on the other genes in the community, but it does not consider the distribution of the genes within the communities. Even if the identified communities do not include genes which are significantly different among samples, SVD still returns the set of genes which have high entropy. Thus, it seems logical and efficient to combine the two methods (t-test and SVD) into a new approach that considers both the whole data and data distribution while ranking.

Dealing with noisy microarray data is challenging. T-test does not consider outliers in the data while ranking because it assumes the data follows normal distribution (and hence each community of genes is well structured), which is not the real case in practice. To take this into account, we apply iterative t-test for determining the p-value for each gene under perturbation by eliminating one sample at a time; we eliminate weak noisy genes by neglecting all genes which do not show significant p-value under all of the conditions. Eliminating more samples is not preferred as the number of samples gets smaller, a case that may lead to information loss; besides, the assumption for t-test may not be satisfied any more.

1.2 Contributions

The basic contributions of this chapter can be enumerated as follows:

1. Iterative t-test has been proposed to minimize the false discovery rate of regular t-test. The p-value of each gene is calculated under perturbation; we consider the removal of one sample at a time.
2. We have modified the SVD based gene selection approach of Varshavsky *et al.* [12] to produce an augmented method useful for handling our problem.
3. A hybrid approach, which combines t-test and SVD, has been proposed for gene extraction from binary and multi-class data.
4. SVM has been used for classification as it showed to be very efficient.
5. The proposed methods have been applied on three microarray datasets. The reported results demonstrate the applicability and effectiveness of the proposed methodology.

The rest of this chapter is organized as follows. The proposed gene selection methods are described in Section 2. The related work is covered in Section 3. Experimental results are reported and discussed in Section 4. Section 5 is summary and conclusions.

2 Alternative Gene Selection Methods

Motivated by the analysis conducted on the existing gene selection methods, we developed some novel approaches for gene selection. Each of the proposed methods integrates different techniques and aspects of the analyzed gene communities into a unique platform. The developed gene selection methods help us to initially filter the genes and then extract the most significant genes within each community and then feed them to the classification method.

2.1 Iterative t-Test

As part of the microarray data preprocessing, it is required to remove noisy genes; these genes are automatically eliminated by the double clustering approach (as described in [8,9,10,11]) because noisy genes are located away from the centroids; therefore they are never selected as cluster representatives. The iterative t-test method described in this section is another way for eliminating noisy genes.

Since there are a variety of systematic errors in microarray and image processing step, some genes show very high expression level under one sample in a class, though the other genes in the same community show low expression level. Such kind of outlier expression level shouldn't affect the gene selection process. Using regular t-test doesn't consider eliminating such outliers; consequently, we have tested how one sample elimination can affect the gene selection process using t-test.

The t-test assesses whether the means of two groups are statistically different from each other. Here, the null hypothesis is expressed as: *the means of genes in the two conditions is equal*; *i.e.* , $H_o : \mu_1 = \mu_2$. Given the replicas of particular treatment and control samples, it is possible to compute the t-test for any gene g to assess if it is differentially expressed by using the following formula under the assumption that genes have differing standard deviations [13],

$$t_g = \frac{\overline{x}_{g,t} - \overline{x}_{g,c}}{\sqrt{\frac{s^2_{g,t}}{n_t} + \frac{s^2_{g,c}}{n_c}}} \tag{1}$$

where $\overline{x}_{g,t}$ and $\overline{x}_{g,c}$ are means of replicas of treatment and control conditions with respective standard deviations $s^2_{g,t}$ and $s^2_{g,c}$, and replica counts n_t and n_c for gene g. It is clear that t-test favors for large mean differences and small standard deviations; it is a good balance between them. The t-test can also be used to find significant genes across more than two conditions.

We applied t-test on the genes to extract those which show significant pattern under perturbation. We apply perturbation by removing samples one by one and find the p-values for the genes under all conditions. We eliminate one gene at a time in order to avoid information loss. When we remove one sample, we find all genes whose p-values are less than a threshold, say 0.001 for example. After that, we generate a matrix called $Significant\ Genes$, where each row contains significant genes under certain condition (removal of one sample), then we find the most significant association rules by considering the frequent set(s) with the maximum support value; surprisingly, the conducted tests reported rules with 100% support.

We sort the frequent sets in descending order by their support value; and then we consider the genes that appear in the rules that have the highest rank. The process applied in this study can be summarized as follows: If the gene has a p-value less than the threshold under all the conditions, then it is significant. After getting the significant genes, we feed them to SVM for classification. We have compared our results with regular t-test, and we have shown that the genes eliminated by our approach are kind of false positives because they have low classification accuracy. To summarize this approach, we eliminate the first sample in the data and we find the p-value for each gene using t-test. Genes with p-value less than the threshold are stored in the first row of $Significant\ Genes$ matrix. Then, we return the first sample and eliminate the second; we find p-values for all genes and store the genes whose p-values are less than the threshold in the second row of the matrix. We repeat the same process for all samples, *i.e.*, at step $i > 1$, we return sample $(i-1)$ and remove sample i. At the end, we take the genes found in every row as significant and not true positives.

2.2 Combining SVD and T-Test for Gene Selection

We have shown that using the SVD-based approach proposed by Varshavsky *et al* [12] is not appropriate for gene extraction from multi-class data. Assume we have a microarray of two classes each having two samples denoted: $Class1_S1$, $Class1_S2$, $Class2_S1$, $Class2_S2$. A gene of values [0,0,1,1] should be significant to distinguish between the two classes. However, the SVD approach by Varshavsky considers a gene whose values are [0,1,1,0] as significant, although it should not be. This inspired us to adapt the SVD approach to two class data in order to extract significant genes. The importance of each gene is computed as in Equation 4.

SVD is a linear transformation of the expression data from n-genes by m-array represented by a matrix $A_{m\times n}$ to the reduced diagonal L-eigengenes by L-eigenarrays matrix, where $L = \min(n, m)$ [14] and $s_i, i =, \ldots L$ are the singular values. Alter *et al* [14] calculated the normalized relative significance p_k of the k-th eigengene for $A_{m\times n}$ as follows:

$$p_k = \frac{s_k^2}{\sum_{i=1}^{L} s_i^2} \tag{2}$$

and the Shannon entropy of the data represented by $A_{m\times n}$ is calculated as:

$$E(A_{m\times n}) = -\frac{1}{log(L)} \sum_{k=1}^{L} p_k log(p_k). \tag{3}$$

Varshavsky *et al* [12] have defined the contribution of the i-th gene CE_i by a leaving out comparison as:

$$CE_i = E(A_{m\times n}) - E(A^{(i)}_{m-1\times n}) \tag{4}$$

where $A^{(i)}_{m-1\times n}$ is the matrix $A_{m\times n}$ with the i-th row deleted.

Genes with high E value are selected as important. In order to adapt the SVD approach to the binary classification problem, we need to compute the average for the values of each gene under each class. This way, we reduce the dimensionality of the data from $m \times n$ to $m \times 2$. Such reduction helps us to identify the genes which have high entropy due to sample difference. The SVD-based approach considers the entropy of the gene with respect to the other genes in the data; and t-test considers the data distribution for each gene. Combining both SVD and t-test will give a better idea about the significance of each gene. To realize this combination, we have defined a new term, denoted $SVDttest$, which is actually computed as the ratio of SVD over t-test:

$$SVDttest(g) = \frac{CE_g}{t_g} \tag{5}$$

where CE_g is computed by Equation 4 and t_g is computed by Equation 1. Based on some thorough testing and analysis of the results, we realized that genes with $SVDttest$ value greater than 1 are considered as significant.

We start with a full microarray data of two classes and many samples in each. We then reduce dimensionality to two by averaging samples in each class. We use Equation 4 to calculate the entropy of each gene. The latter shows how the entropy of the matrix is affected when the gene is removed. The case where the entropy does not change indicates that the gene is not important. On the other hand, the significance of the gene increases as the change in the entropy increases. The advantage of reducing the dimensionality of the genes into 2 ensures that genes with high entropy result from the difference across samples. Unlike the method proposed by Varshavsky *et al* which does not ensure that high entropy genes are due to the difference in classes. It just ensures that the genes have dynamic gene expression profiles along samples, but not necessarily across classes.

The new approach proposed in this chapter considers both statistical and entropy based significance for each gene. Our approach is summarized as follows:

1. Find the p-value of each gene using t-test
2. For each gene, average the gene expression value under each condition, *i.e.*, if we have two class data, then we end up having $N \times 2$ data, where N is the number of genes.
3. Find the contribution of each gene to the entropy of the matrix using SVD as in Equation 4.
4. For each gene, divide the entropy contribution calculated in step 3 by the p-value from step 1, and select genes with $SVDttest$ value greater than 1. The conducted tests demonstrate that the larger the score, the more significant the gene is.

For a gene to have high $SVDttest$ value, it should have either very large SVD value which results from the difference across classes, or very small p-value which results from large difference across classes. The conducted tests demonstrate that most genes with high $SVDttest$ also showed to have very small p-value, although the order of the genes in each approach was different.

3 Related Work

Golub *et al* [5] may be considered as the first group who tried to distinguish between AML and ALL based on gene expression data. They used a model of SOM in combination with a weighted voting scheme. They obtained a strong prediction of 29/34 samples in the test data using 50 genes. Furey *et al* [15] applied SVMs to the AML/ALL data. Significant genes are derived from a score calculated from the mean and standard deviation of each gene type. Tests are performed for 25, 250, 500, and 1000 top ranked genes. At least two test examples are misclassified in all SVM tests. Guyon *et al* [16] also applied SVM, but with a different feature selection method called recursive feature elimination. For each iteration, the weights associated with genes are recalculated by a SVM test, and genes with the smallest weights are removed. On 8 and 16 genes, the classification error on the test set is zero.

Li and Wong [17] used a new feature selection method called emerging patterns. When they applied their method on the AML/ALL data, they were able to identify one gene (zyxin), which was able to classify 31/34 of the samples. Toure and Basu [7] applied neural network to cancer classification; 10 genes were used for classification purposes. The neural network was able to fully separate the two classes during the training phase. However, the classification of the test set samples did not achieve high accuracy, 15 samples were misclassified. Zhang and Ke have used the 50 genes from the paper of Gloub *et al* and applied SVM and CSVM for classification [18]. Two misclassifications occurred while using SVM, but no errors were reported when CSVM was used. In another work, Leping *et al* [19] applied GA/KNN method to the same data set; 50 genes were used for classification; GA/KNN correctly classified all training set samples, and all but one of the 34 test samples. Bijlani *et al* [20] used independently consistent expression discriminator (ICED) for feature extraction. They could select 3 AML distinctors and 13 ALL distinctors, which were able to classify the training test without any errors; but one sample was misclassified in the test data.

As the colon dataset is concerned, Bicciato *et al* [21] used Autoassociative Neural network model for classification. No sample was misclassified during the training session, while seven samples received wrong classification during the testing phase. In another study in which Moler *et al* [22] used SVM for classification, 4 samples were misclassified. Wang *et al* [23] used two novel classification models. The first was a combination of optimally selected self-organizing map (SOM), followed by fuzzy C-means clustering. And the second was the use of pair-wise Fisher's linear discriminant. In the former model 12% of the samples were misclassified and in the latter model, 18% of the samples were incorrectly classified. At least 4 misclassifications occurred in most of the studies carried out.

Recently, Lipo *et al* [24] argued that they could find the smallest number of genes which are able to correctly classify samples. They have used t-test and class separability methods for removing non-significant genes, and SVM and Fuzzy neural networks for classification. In order to find the best genes, they used all the filtered genes from t-test individually to classify the data. If the accuracy is low, they use all possible binary combinations. As long as the accuracy is low, they keep combining genes until the accuracy becomes high. We argue that this approach is not reliable at all as it has several drawbacks. First, they don't consider the false discovery rate while selecting the p-value threshold. This approach has neither mathematical nor biological significance. They try all genes individually first, then they try to find all combinations which may give good classification accuracy. If the two classes were very close together, their method will not give good classification; thus, can't find best genes. Also, if a set of genes could individually have high classification, the following question would arise, which one to select? we think this approach is data dependent, classification method dependent, and the genes extracted from different data may have different characteristics. In another study, Lipo and Feng [25] have shown the applicability of SVM to correctly classify microarray data.

A novel approach of using SVD for feature extraction was proposed by Varshavsky *et al.* [12]. This approach was used to select feature, which cause high entropy in the data, for clustering purposes not for classification. The approach is based on leave-one-out comparison. In this chapter, we have modified this approach in order to adapt it to our classification problem. Actually, multiple ordering criteria showed to be very effective in selecting significant genes. On the other hand, using single selection methods may have some drawbacks which may be eliminated when the method is used in combination with others. A three layer ranking algorithm was proposed by Chen *et al.* [26]. They have combined fold change, t-test and SVM-RFE for gene selection. Volcano plot was used to select the genes with high rank in both methods.

Redundancy is another criteria that should be optimized while selecting genes for cancer classification. Most of the existing approaches, namely t-test, SAM, SVM-RFE, SVD, etc, don't consider this point. Using clustering for redundancy elimination is effective for microarray data. Clustering genes will group those with similar functionality in one cluster. Thus, by selecting representative features from each cluster, we ensure that all of the selected genes have different functions. We have used double clustering to find representative genes in each cluster [8]. The results we got are very promising. Another approach which is very close to ours applies clustering to reduce redundancy [27]. They first cluster the data set into 20 clusters using FCM, then select the top genes from each cluster which show significant pattern along the samples using t-test. They selected at least one gene from each cluster. For bad clusters, they selected more than one gene such that the correlation between the selected genes is minimum. Highly correlated genes are shown to have similar functions and exist in the same cellular pathway. Finally, by considering the work being done by other members of our research group, we realized that applying association rule mining to extract significant genes showed to be a promising method [11]. Benefiting from this experience, we have developed fuzzy association rule based gene extraction method [10] which considers correlation among genes in the extraction process. Using the training set, we first extract all possible rules after discretizing the data; survival rule will be used for classification. However, the details of this work are out of the scope of this thesis.

As a closing remark, the observation we have drawn from our study described in this section could be articulated as follows. Working out a solution for a given problem might lead to some valid results. However, combining multiple aspects and perspectives into the solution produces more satisfactory and relevant results.

4 Experimental Analysis

As the methodology developed in this chapter is concerned, we decided to run a set of experiments using different datasets that will allow us to highlight the strengths of our proposed methods as compared to the similar methods from the literature. In this section, we discuss the conducted experiments. We highlight the results of our approaches and evaluate their performance and applicability. We report the results from the comparison of our approach with other existing methods which investigated the same cancer classification problem. We have also analyzed the results both in terms of classification accuracy and biological significance.

4.1 Testing Environment

Data preprocessing has been conducted using matlab 7.0. Clustering and the cluster validity analysis have been conducted using the clustering package available within Matlab. Gene selection has been performed using t2test in matlab. For classification, we have used LIBSVM package implemented in matlab. LIBSVM is a free library for classification and regression available online at (`http://www.csie.ntu.edu.tw/~cjlin/libsvm/`). The SVD based gene selection code was kindly provided by Varshavsky *et al.* [12]. We have run the experiments on laptop with Intel Core2Duo CPU 2.0GHz and 1.99GB of RAM, running Windows XP professional version 2002 SP2.

4.2 The Data Sets

In this work we have used three cancer data sets:

1. Acute myeloid leukemia (AML)/Acute lymphoblastic leukemia (ALL) taken from [5],
2. Colon data set from [28], and
3. Breast cancer data taken from [29].

A brief summary of the datasets is given next. The AML/ALL data contains 7,130 gene and 73 patient samples; 38 sample for training, 27 AML / 11 ALL, and 35 for testing, 23 AML / 12 ALL. As a normalization step, the intensity values have been normalized such that the overall intensities for each chip are equivalent. This is done by fitting a linear regression model using the intensities of all genes in both the first sample (baseline) and each of the other samples. The inverse of the "slope" of the linear regression line becomes the (multiplicative) re-scaling factor for the current sample. This is done for every chip (sample) in the data set, except the baseline which gets a re-scaling factor of one. A missing value means that the spot was not identified. The missing values were predicted according to the neighbor values. The colon cancer data has 62 patient samples; 40 tumor and 22 normal; 2,000 genes were studied. Samples were split as follows, 15 normal samples were used for training and 7 for testing; 23 tumor samples were used for training and the other 17 were used for testing. For breast data, it has 7,129 genes and 47 samples. Samples were 23 estrogen receptor positive split as 15 for training and 8 for testing, and 24 estrogen receptor negative split as 15 for training and 9 for testing. Two forms of the data were used; normal data and log data by using log function for each value.

In the rest of this section, we demonstrate the results of the proposed methodologies. We first show the results of the iterative t-test as accuracy of the extracted gene set and the accuracy of the eliminated gene set. Then, we present the results of the SVDttest and highlight the outperforming nature of the proposed method.

4.3 Iterative Test Based Gene Reduction Results

Employing t-test for identifying differentially expressed genes is a well used technique as it is easy to apply, but setting the threshold p-value is critical; 0.05 or 0.001 are the most used p-values. Selecting threshold p-value can have many consequences, like high false positive and false negative rates. Bonferroni correction can be used to set critical p-value. It considers the number of times the test is used. Basically, the critical $p - value = \alpha N$, where N is the number of samples. But this method doesn't discover the real false positives. In here, we have applied the proposed approach for AML/ALL data in *log* form. First, we applied t-test and selected all the genes having p-value less than 0.001; as a result, 40 genes were selected. We used the 40 genes as significant genes to be fed to SVM. Results of the classification are shown in Table 1.

After applying the iterative t-test, we reduced the number of genes to 25. We used both the selected 25 genes and the eliminated 15 genes for classification. The results are shown in Table 2 and Table 3.

We observed that the 25 genes are efficient to classify AML and ALL samples. The 15 eliminated genes showed to give very poor accuracy; they are not significant within their social communities. This indicates that the eliminated genes are more likely to be false positives. Most of the selected genes are cancer related as they showed to be related to histone complexes, cell cycle and

Table 1. Classification results of the 40 genes selected using t-test of p-value 0.001

	Linear SVM	**Polynomial SVM**	**RBF SVM**
Accuracy	*94%*	*94%*	*94%*
Cross-validation	*100%*	*97%*	*97%*

Table 2. Classification results of the 25 genes selected using iterative t-test

	Linear SVM	Polynomial SVM	RBF SVM
Accuracy	*94%*	*94%*	*94%*
Cross-validation	*100%*	*100%*	*100%*

Table 3. Classification results of the 15 eliminated using iterative t-test

	Linear SVM	Polynomial SVM	RBF SVM
Accuracy	*58%*	*61%*	*58%*
Cross-validation	*100%*	*97%*	*100%*

oncogenes. The 25 genes selected using our iterative approach are key players within their communities; they are listed in Table 4, where the last column indicates how other researchers have already identified these genes as significant.

4.4 SVD-t-Test Based Gene Selection Results

We have applied our proposed hybrid approach on the three data sets described in Section 4.2. We studied the effectiveness of the selected genes for classification. Here we show the accuracy of the classifiers when fed with the selected genes and we highlight the advantages of the hybrid approach compared to t-test and adapted SVD.

AML/ALL Data. Applying the hybrid approach to select differentially expressed genes from AML/ALL data resulted in selecting 13 genes as significant in their communities. Using these genes, we got the results reported in Table 5. Only one sample; number 70, was misclassified.

The main advantage of svd-ttest is that it is easier to make the cutoff value as shown in Table 6. Also, interestingly we realized that the 13 top genes are among the 25 genes selected by the iterative t-test. This indicates that the proposed approach is efficient for false discoveries elimination and it is equivalent to the iterative t-test with the advantage that SVD-ttest considers the correlation among genes. Here, it is worth mentioning that the higher the SVD-ttest or SVD value, the more significant the genes is within its social community and hence more relevant to consider further in the analysis. The SVD-ttest values are widely distributed and follow a normal distribution, unlike SVD and t-test values.

Colon Data. We have applied the hybrid approach on the colon data. As a result, we have selected 11 genes as significant within their communities. The results of the classification are shown in Table 7. Using linear and RBF SVM, 17 out of 24 genes were correctly classified; however, using polynomial SVM 19 out of 24 were correctly assigned to the right class.

Table 4. Top 25 genes selected by iterative t-test

Gene index in AML/ALL data	Gene Description	Gene Accession Number	Literature Evidence
532	*HMG1 High-mobility group protein 1*	*D63874-at*	
804	*Macmarcks*	*HG1612-HT1612-at*	
1394	*FTH1 Ferritin heavy chain*	*L20941-at*	
1674	*FTL Ferritin, light polypeptide*	*M11147-at*	
1704	*ADA Adenosine deaminase*	*M13792-at*	*[5]*
1897	*HNRPA2B1 Heterogeneous nuclear ribonucleoprotein A2/B1*	*M29064-at*	
1928	*Oncoprotein 18 (Op18) gene*	*M31303-rna1-at*	*[5, 21]*
2111	*ATP6C Vacuolar H+ ATPase proton channel subunit*	*M62762-at*	*[5]*
2121	*CTSD Cathepsin D (lysosomal aspartyl protease)*	*M63138-at*	*[5, 21]*
2186	*MAJOR HISTOCOMPATIBILITY COMPLEX ENHANCER-BINDING PROTEIN MAD3*	*M69043-at*	
2354	*CCND3 Cyclin D3*	*M92287-at*	*[5, 21]*
2394	*PLCB2 Phospholipase C, beta 2*	*M95678-at*	
3258	*Phosphotyrosine independent ligand p62 for the Lck SH2 domain mRNA*	*U46751-at*	*[5]*
3320	*Leukotriene C4 synthase (LTC4S) gene*	*U50136-rna1-at*	
4211	*VIL2 Villin 2 (ezrin)*	*X51521-at*	
4328	*PROTEASOME IOTA CHAIN*	*X59417-at*	*[5, 21]*
4792	*SERYL-TRNA SYNTHETASE*	*X91257-at*	
5122	*GB DEF = CD36 gene exon 15*	*Z32765-at*	
5501	*TOP2B Topoisomerase (DNA) II beta (180kD)*	*Z15115-at*	
5890	*Dematin*	*HG4535-HT4940-s-at*	
5899	*Rhesus (Rh) Blood Group System Ce-Antigen, Alt. Splice 2, Rhvi*	*HG627-HT5097-s-at*	
6247	*PIM1 Pim-1 oncogene*	*M54915-s-at*	
6561	*Metargidin precursor mRNA*	*U41767-s-at*	
6806	*Lysozyme gene (EC 3.2.1.17)*	*X14008-rna1-f-at*	
6940	*GB DEF = Chloride channel (putative) 2163bp*	*Z30644-at*	

Table 5. Classification results of the top 13 genes from AML/ALL data using the hybrid approach

	Linear SVM	**Polynomial SVM**	**RBF SVM**
Accuracy	*97%*	*97 %*	*97%*
Cross-validation	*92.1%*	*92.1%*	*97.3%*

Table 6. Comparison among t-test, SVD, and SVD-ttest cutoff values

Gene index	SVD-ttest	t-test	SVD
3320	*4.3974e+005*	*1.1077e-010*	*4.87e-05*
2121	*347.5*	*1.9935e-007*	*6.93e-05*
6806	*152.33*	*7.813e-007*	*0.000119*
3258	*101.03*	*7.1392e-007*	*7.21e-05*
804	*13.642*	*4.0577e-006*	*5.54e-05*
2111	*13.18*	*3.2612e-006*	*4.30e-05*
2186	*10.737*	*9.9251e-006*	*0.000107*
5501	*10.184*	*6.3121e-006*	*6.43e-05*
4328	*9.4266*	*8.6175e-006*	*8.12e-05*
1928	*9.3483*	*3.5036e-006*	*3.28e-05*
4211	*4.0965*	*1.0748e-005*	*4.40e-05*
1673	*2.9822*	*1.0747e-005*	*3.21e-05*
1704	*2.2168*	*3.8127e-005*	*8.45e-05*

Table 7. Classification results of the top 11 genes from colon data using the hybrid approach

	Linear SVM	Polynomial SVM	RBF SVM
Accuracy	*70%*	*79 %*	*70%*
Cross-validation	*92.1%*	*97.3%*	*94.7%*

Table 8. Classification results of the top 12 genes from the breast data using the hybrid approach

	Linear SVM	Polynomial SVM	RBF SVM
Accuracy	*88%*	*94%*	*94%*
Cross-validation	*86.6%*	*90%*	*90%*

Table 9. Classification results of the top 3 genes from the breast data using the hybrid approach

	Linear SVM	Polynomial SVM	RBF SVM
Accuracy	*94%*	*94%*	*94%*
Cross-validation	*90%*	*90%*	*90%*

Breast Data. We have also applied the hybrid approach on the breast data; we have selected the top 3 genes and we got the results reported in Table 8. Using linear SVM, 2 samples out of 17 were misclassified and only one was misclassified using polynomial and RBF SVM. We obtained even the best results using only the top 3 genes as indicated in Table 9.

5 Discussion and Conclusion

The target of this work is to extract representative features from social communities of genes. As part of microarray data preprocessing, significant gene selection is crucial for better and accurate classification. To cope with this, we proposed methods capable of successfully identifying significant genes within the communities. The first method is capable of handling the noisy data, and hence eliminates genes which are noisy. The second method eliminates the genes which don't show high entropy and statistical significance. As a result, the proposed approaches significantly reduce the false discovery rate. The conducted tests demonstrate the significance of the proposed approaches as interesting contribution for more appropriate gene selection.

As the other methods are concerned, the literature shows that t-test has been widely used for gene selection, but choosing the threshold is very critical and has absolute boundary. For example, setting the threshold to be 0.01 will lead to selecting a gene which has p-value of 0.009999 and exclude a gene which has p-value as 0.0101. Furthermore, we may select all the genes in the data if all of them demonstrate to be statistically significant among samples. The reason behind this is the lack of the ability to consider the whole data while selecting the genes. On the other hand, the SVD-based approach proposed by Varshavsky *et al* [12] does consider the whole data while selecting the genes; however, it still selects a set of genes even if no gene in the set is statistically significant. The idea of SVD-ttest has been inspired from the drawbacks of those two methods. Proposing a method which can consider the statistical significant of the individual genes within each community and their entropy on the whole social network is very important. Another advantage of the proposed approach is that there is no need for a cutoff value. Statistically significant genes with large entropy are selected. There still does not exist a solid interpretation behind this, but experimentally it showed to be working very well. Analyzing the biological importance of the selected genes, we have seen that they participate in variant processes in the cell like cellular iron ion homeostasis, cell differentiation, proteolysis and T-cell activation and cell-cell adhesion. In addition, the genes selected from AML/ALL do have role in apoptosis, leukotriene biosynthesis, ubiquitine -dependent protein catabolic process, and inflammatory responses in addition to cytoskeletal anchoring. This shows that the genes selected by each of the proposed methods don't have common functionality among them and they do represent most of the cellular functionalities related to cancer cells. However, the the two methods reported genes with similar functionality because the degree of significance of different genes within their communities is considered differently by the different methods. Interestingly, we have see that two of the selected genes are involved in iron transport, which makes the iron transport process a target for more investigation about the exact role of iron in AML or ALL. Currently we are expanding this work into a more robust automated approach that integrates all the perspectives under one umbrella where the genes reported by the different methods will be analyzed in a collaborative manner to select the most optimal set of biomarkers.

References

1. Jemal, A., Siegel, R., Ward, E., Xu, T., Thun, M.J.: Cancer statistics. A Cancer Journal for Clinicians 57, 43–66 (2007)
2. Butte, A.: The use and analysis of microarray data. Nature Reviews 1, 951–960 (2002)
3. Dembele, D., Kastner, P.: Fuzzy c-means method for clustering microarray data. Bioinformatics 19, 973–980 (2003)
4. Kohonen, T.: Self-organizing paps. Springer Series in Information Sciences, vol. 30. Springer, Heidelberg (2001)
5. Golub, T.R., Slonim, D., Tamayo, P., Huard, C., Gaasenbeek, M., Mesirov, J., Coller, H., Loh, M., Downing, J., Caligiuri, M., Bloomfield, C., Lender, E.: Molecular classification of cancer: class discovery and class prediction by gene expression monitoring. Science 286, 531–537 (1999)
6. Hsu, C., Chang, C., Lin, C.: A practical guide to support vector classification. Technical report, Department of Computer Science, National Taiwan University (July 2003)
7. Toura, A., Basu, M.: Application of neural network to gene expression data for cancer classification. In: Proceedings of IEEE International Joint Conference on Neural Networks, pp. 583–587 (2001)
8. Alshalalfa, M., Alhajj, R.: Application of double clustering to gene expression data for class prediction. In: Proceedings of AINA Wokshops, vol. 1, pp. 733–736 (2007)
9. Alshalalf, M., Alhajj, R.: Attractive feature reduction approach for colon data classification. In: Proceedings of AINA Workshops, vol. 1, pp. 678–683 (2007)
10. Khabbaz, M., Kianmher, K., Alshalalfa, M., Alhajj, R.: Fuzzy classifier based feature reduction for better gene selection. In: Song, I.-Y., Eder, J., Nguyen, T.M. (eds.) DaWaK 2007. LNCS, vol. 4654, pp. 334–344. Springer, Heidelberg (2007)
11. Kianmehr, K., Alshalalfa, M., Alhajj, R.: Effectiveness of fuzzy discretization for class association rule-based classification. In: Proceedings of the International Symposium on Methodologies for Intelligent Systems. LNCS. Springer, Heidelberg (2008)
12. Varshavsky, R., Gottlieb, A., Linial, L., Horn, D.: Novel unsupervised feature filtering of biological data. Bioinformatics 22, 507–513 (2006)
13. Dudoit, S., Yang, Y.H., Callow, M., Speed, T.: Statistical methods for identifying differentiallyexpressed genes in replicated cdna microarray experiments. Technical Report #578, University of California, Berkeley (2000)
14. Alter, O., Brown, P., Botstein, D.: Singular value decomposition for genome-wide expression data processing and modeling. PNAS 97, 10101–10106 (2000)
15. Furey, T.S., Cristianini, N., Duffy, N., Bednarski, D.W., Schummer, M., Haussler, D.: Support vector machine classification and validation of cancer tissue samples using microarray expression data. Bioinformatics 16, 906–914 (2000)
16. Guyon, I., Weston, J., Barnhill, S., Vapnik, V.: Gene selection for cancer classification using support vector machines. Machine Learning 46, 389–422 (2002)
17. Li, J., Wong, L.: Identifying good diagnosis gene group from gene expression profile using the concept of emerging patterns. Bioinformatics 18, 725–734 (2002)
18. Zhang, X., Ke, H.: All/aml cancer classification by gene expression data using svm and csvm. Genomics informatics 11, 237–239 (2000)
19. Li, L., Pedersen, L.G., Darden, T.A., Weinberg, C.R.: Class prediction and discovery based on gene expression data. Iostatistics Branch and Lab of Structural Biology, National Institute of Environmental Health Sciences, Research Triangle Park, North Carolina (2000)

20. Bijlani, R., Cheng, Y., Pearce, D., Brooks, A., Ogihara, M.: Prediction of biologically significant components from microarray data: independently consisitent expression discriminator(iced). Bioinformatics 19, 62–70 (2003)
21. Bicciato, S., Pandin, M., Didon, G., Bello, C.D.: Pattern identification and classification in gene expression data using an autoassociative neural network model. Biotechnology and Bioengineering 81, 594–606 (2002)
22. Moler, E., Chow, M., Mian, I.: Analysis of molecular profile data using generative and disciminative methods. Physiol. genomics 4, 109–126 (2000)
23. Wang, J., Hellem, T., Jonassen, I., Myklebost, O., Hovig, E.: Tumor classification and marker gene prediction by feature selection and fuzzy c-means clustering using microarray data. BMC Bioinformatics 4, 60–70 (2003)
24. Wang, L., Chu, F., Xie, W.: Accurate cancer calssification using expressions of very few genes. IEEE/ACM transactions on computational biology and bioinformatics 4, 40–53 (2007)
25. Chu, F., Wang, L.: Cancer classification with microarray data using support vector machines. Bioinformatics using computational intelligence paradigms 176, 167–189 (2005)
26. Chen, J., Tsai, C., Tzeng, S., Chen, C.: Gene selection with multiple ordering criteria. BMC Bioinformatics 8 (2007)
27. Jaeger, J., Sengupta, R., Ruzzo, W.L.: Improved gene selection for classification of microarrays. In: Proceedings of Pacific Symposium on Biocomputing, pp. 53–64 (2003)
28. Alon, U., Barkai, N., Notterman, D.A., Gish, K., Ybarra, S., Mack, D., Leving, A.: Broad patterns of gene expression revealed by clustering analysis of tumor and normal colon tissues probed by oligonucleotide array. PNAS 96, 6745–6750 (1999)
29. West, M., Dressman, H., Haung, E., Ishida, S., Spang, R., Zuzan, H., Olson, J., Marks, J., Nevins, J.R.: Predicting the clinical status of human breast cancer by using gene expression profiles. PNAS 98, 11562–11567 (2001)

Author Index

MIX
Papier aus verantwortungsvollen Quellen
Paper from responsible sources
FSC® C105338

If you have any concerns about our products,
you can contact us on
ProductSafety@springernature.com

In case Publisher is established outside the EU,
the EU authorized representative is:
Springer Nature Customer Service Center GmbH
Europaplatz 3, 69115 Heidelberg, Germany

Printed by Libri Plureos GmbH
in Hamburg, Germany